Technology as Institutionally Related to Human Values

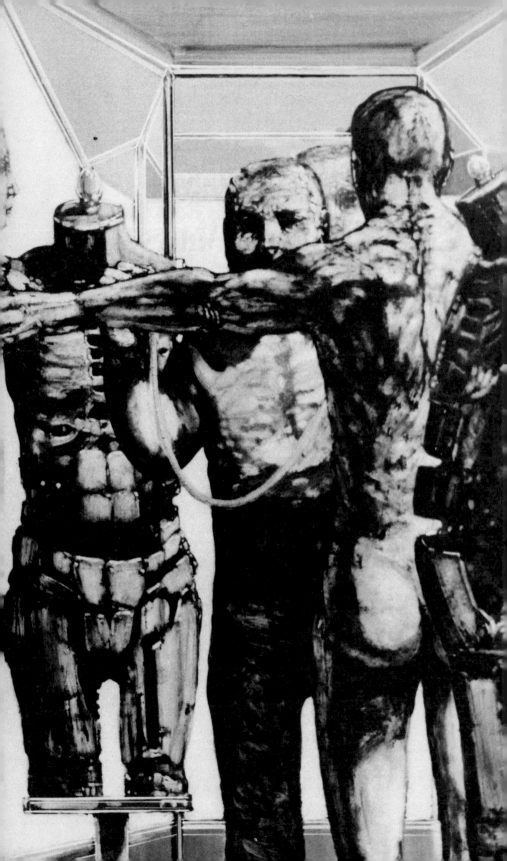

The "Prometheus" Series on Contemporary Institutions

Technology as Institutionally Related to Human Values

Edited by
PHILIP C. RITTERBUSH

Assisted by
Martin Green

ACROPOLIS
BOOKS FOR
Contemporary
EDUCATION

Published by **ACROPOLIS BOOKS LTD.**, Washington, D.C. 20009

PROMETHEUS: STUDIES OF CONTEMPORARY INSTITUTIONS

Second Series, number three (February, 1974)

LET THE ENTIRE COMMUNITY BECOME OUR UNIVERSITY
(number one, April, 1973)

NATIONAL REGISTER OF INTERNSHIPS AND EXPERIENTIAL EDUCATION
(number two, September 1973)

Published in cooperation with
THE ARCHIVES OF INSTITUTIONAL CHANGE
Wallpack Village, New Jersey 07881.

Editor: Philip C. Ritterbush
Review Editor: Deborah Goldman
Art Editor: Lenore F. Sams (Mrs.)

ACROPOLIS BOOKS LTD.
Colortone Building, 2400 17th Street, N.W.
Washington, D.C. 20009

Printed in the United States of America by
COLORTONE PRESS, Creative Graphics Inc.
Washington, D.C. 20009

Library of Congress Number 74-320
International Standard Book Number 87491-511-2 cloth
87491-512-0 paper
International Standard Serial Number 0048-5535

The assistance of the Rockefeller Brothers Fund
and the U.S. National Park Service in the completion of this book
is gratefully acknowledged. Belinda Barrington prepared
the bibliography and institutional summaries.
Designed by Bill Meadows.

Contents

Some Humanistic Interactions for   *Martin Green*
Technological Education   9

Colloquium: The Role of the
Humanities at M.I.T.

    The Function of the   *Richard Douglas*
    Humanities at M.I.T.   15

    Engineers and Humanists   *David J. Rose*   19

    A Bridge: The History of   *Brooke Hindle*
    Technology   24

    Some Thoughts about the   *Judith Wechsler*
    Humanities at M.I.T.   28

    The Integration of Knowledge: *Christopher Schaefer*
    Notes on Curriculum Change
    at M.I.T.   32

    Improving M.I.T. as an   *Commission on M.I.T.*
    Environment for Humane   *Education*
    Learning   39

Future Directions for   *Vary T. Coates*
Technology Assessment   47

Technology Assessment: *Michael Gibbons*
Objectivity or Utopia?   59

James K. Feibleman's View of   *Stanley C. Feldman*
Technology and the Problem
of Ideologies   67

Using Technology to Change   Brewster C. Denny
Society: A Political Problem   71

Some Tasks for "The Humanities"   Max Black   83

Military Technology: A Problem   John Kenneth Galbraith
of Control   94

Technology and New Institutions   John P. Eberhard
in Human Settlements   101

The Craftsman: Some Reflections   David E. Whisnant
on Work in America   109

Art, Technology, and Ideology: The   John Adkins Richardson
Bauhaus as Technocratic Metaphor   127

Institutions Where Technology   Belinda Barrington
and Human Values Meet   142

Technology as an Institution: Belinda Barrington and
Social and Cultural Aspects   Philip C. Ritterbush
(Bibliography)   157

Index   191

ILLUSTRATIONS

Vincent Perez, born in New Jersey in 1938, resides in Alameda, California and exhibits his work at the William Sawyer Gallery in San Francisco. His work has been shown in numerous exhibitions including the National Portrait Gallery, the Salt Lake City Art Center, and the LaJolla Museum of Art. He is Associate Professor of Art at the California College of Arts and Crafts, Oakland. The works reproduced in this volume exemplify his concern for the transmutations of man and his values in our technological world.

	Page
"Tailor Shop" panel	*frontispiece*
Illustration, untitled, from the "the Art and the Tools," Searle Medidata, Inc.	8
"Le Coq" (woodcut)	21
"Birdbox" (woodcut)	31
"Man Thinking" (woodcut)	34
"Horseman" (woodcut)	48
"Sleep" (woodcut)	70
"The Barber" (woodcut)	78
"Installed Figures" (woodcut)	91
"Car Door #3" closed (acrylic)	103
"Car Door #3" open (acrylic)	104
"Musicians" (acrylic)	118
"Lockers" (acrylic)	129
"Mirror" (acrylic tryptich)	138

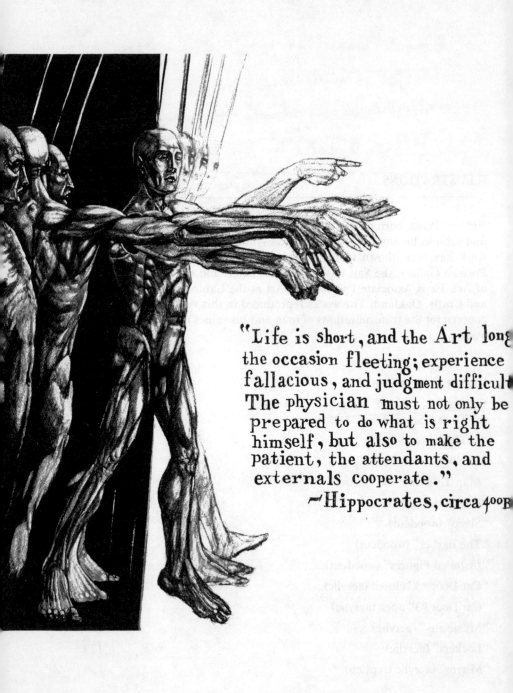

"Life is short, and the Art long,
the occasion fleeting; experience
fallacious, and judgment difficult.
The physician must not only be
prepared to do what is right
himself, but also to make the
patient, the attendants, and
externals cooperate."

—Hippocrates, circa 400 B.

Some Humanistic Interactions for Technological Education

Martin Green
Professor of English
Tufts University

The problems with which this book grapples are obvious and enormous, but in a sense impalpable. They have so many forms that they seem formless, especially as one watches each new expert seize hold of them by a different aspect. I myself must do just that, of course, but I must also try not to add to that discouraging shapelessness and impalpability. So I want simply to describe the kind of course I would like to see institutes of technology teach. (I shall borrow some ideas from my *Science and the Shabby Curate of Poetry,* of 1964.) I hope that by specifying thirteen such courses, I can create an image of the interaction we all want between technological and humanistic education. If we have a fairly specific image — instead of a number of abstract definitions — we can, I hope, discuss the means to take towards that end in a more fruitful way. Once we know the new kind of course there will be, then the new department to teach it, the new training for its teachers, the new relations with the departments of science and technology, will come easily enough. Many of these mechanical arrangements, after all, do not deserve theoretical discussion. They are to be solved practically, or not at all. When the atmosphere is right, they resolve themselves, and the atmosphere should be the emanation of the shared vision, the image.

I will describe a few courses which could be called science-and-history, and a few that could be called science-and-literature. For general use such categories are unfortunate, because each course will only succeed if it seems to be a unity, not an amalgam. But for us here those categories have the point that they lead the mind on to other groups, to be labelled science-and-music, science-and-art, and so on. There is just as much to be done in those directions as in these, but I am more at ease with this subject-matter.

First then, science-and-history. I would like to see a course, or courses, in the organization of science. For instance, it would be interesting to study how joint or group discoveries were made, and how much play the non-scientific factors had in determining the success or failure of particular enterprises. The financial backing, the enthusiasm of the group, the temperaments of the individuals, the moral issues involved, each is likely to bear at some point on the scientific success achieved. James R. Watson's *The Double Helix* tells us something of that, in the case of the DNA discovery. Let the course's focus be clearly on the scientific success achieved, and how it was achieved, and I think the serious attention of even the purest scientists would be held.

Then it would also be interesting to study how a laboratory director operates, and just what makes a man a great laboratory director. By what means are a group of scientists induced to give of their best work, and in some sense better than their best — better than they ever do in other phases of their careers? This would seem to call for a series of historical case-studies — Rutherford would be a very good case — though one could also consult witnesses from contemporary and on-going enterprises. The obvious method would be to assemble various personal reminiscences, and various pieces of documentary evidence, and to piece together a picture of the individual director for diagnosis.

Finally — within this group of the organization of science — one might focus on the process by which a project transforms itself and becomes successful when it passes from the control of one man or group to another. In all these cases, clearly, the lecturer must not be afraid to talk in detail and frankly about the personalities of the people involved. It is in some sense the whole point of such courses that he should give personal and novelistic accounts of people, with genealogical, sociological, psychological, political detail. He will also have to give some account of institutions, and the conditions that affect them, and other sorts of non-scientific matter. But if the focus is clearly and closely on the functioning of these personalities and institutions as part of a scientific project, on the interaction of idiosyncrasy with talent and work-style, then the interest will not be gossipy.

Then I would like to see a course about each of those striking coincidences in the history of science, rooting them in the other aspects of the history of their times. For instance, the invention of non-Euclidean geometry by Gauss, Lobachevsky, and Bolyai, each working independently, and all within a decade or so. Or the case of Caspar Wessel and Jean-Robert Argand, both of whom worked out the beautiful geometrical interpretation of imaginary numbers, in total ignorance of each other, and one seven years after the other. There is a historical drama implicit in each case; so many centuries without this idea, these minds living without contact with each other, and without much similarity as individuals, but living in response to the same historical forces, and gradually growing into the same activity, and finally making the same discovery. These are the dramas, it seems to me, to arrest the attention of students of science and technology, and to develop their powers of imagination humanistically.

Similarly we could have a course about each of those great disputations over terms; for instance, the debate over 'power', 'force', and 'energy', in the mid-nineteenth century, between Carnot, Joule, Thompson, and Helmholtz. (It would be interesting to keep in view the likenesses and differences between that debate and the quarrel over 'classic' and 'romantic'.) To see the differences in personal and national temperaments affecting the development of science — over the years to come as well as at the moment — stretches the students' minds, by stretching the meaning of 'science' itself.

And then there could be a course on the limits of the different sciences, and the changes of those limits with the passing of time. There is the advance of the idea of mechanics from Galileo to Newton to Boyle, and so on, supplanting with its explanations those earlier theories which had kept the stars, or the earth, or the planets, parts of some quite different branch of knowledge. To which must be added the retreats of that idea from areas it was not, after all, competent to explain; the retreat from Descartes' version of the human body as a crude mechanical system, for example, with nerve tubes filled with compressible fluid. These to-and-fro shiftings of ideas are very vivid when the scientific examples are combined with the non-scientific. (Magnus Pyke has a book on this subject.)

A course in alchemy could lead science students to a confrontation with the imagination in all its complexity. That queer fusion of the scientific with the non-scientific, the magical; and the likeness as well as the unlikeness of some alchemical ideas to those of modern science, illuminates the special character of science very sharply. With the queer fusion of the magical-scientific with the religious and the aesthetic, and the way the whole occult tradition has fertilized the minds of poets, philosophers, psychologists, right up to now, this makes material for several courses. One could, for instance, devote a course to Goethe as an alchemist.

But perhaps we need most of all courses that concentrate on the intellectual personalities of great scientists of the past. What needs to be shown is how a given psychological factor (within the total complex) is related to (expresses itself in, is encouraged by, is in conflict with) the kind of scientific work done. F. W. Aston, we are told, was a neat, prim bachelor, devoted to his hobbies, who early became what is called old-maidish in his general way of life. And his scientific work consisted of finer and finer measurements in the mass-spectograph he designed and built himself. Nearly all his scientific life was passed in one room of the Cavendish, where, alone, without assistants or colleagues, he designed and constructed and used his machine, took it to pieces and built another, and then improved it again. This is not 'just science' nor even just physics; it is a certain kind of physics; which he chose to do. His scientific achievement is just as closely related to his nature, his psychological endowment and what he made of that, as, for instance, Gray's or Pope's literary achievement was related to their endowments.

Michelson, to take another example, was also a measurer and machine-builder, but of a much harsher, more arrogant and domineering, more aggressively masculine type. His scientific work corresponded to his character, and so he built machines very unlike Aston's. Pasteur, again very different, was a kindly, idealistic civil-servant type, devoted to his parents, his children, his people, his country; and (obviously in some sense 'therefore') he spent *his* scientific life on problems of preventing disease and controlling infection and improving industry.

The best book that has treated science in this way is Koestler's biography of Kepler. One might think that that book succeeds where one or another scientist wouldn't, because Kepler was such a rich, eccentric, poetic, mind. But once he starts looking at opposite types of scientist, one sees that they all cry out for similar treatment. Newton is a fair example of an opposite type; but how much there is to be explained temperamentally, in those two bursts of fantastically concentrated work on gravitation, the almost equally fantastic 18 year lapse in between, the fierce resentment and evasion of possible criticism, the frigid refusal of equal relationship or cooperation with anyone. There is an extraordinary temperament there, both very direct and very evasive, and it affected the work done, in choice of subject, relationship to other workers, manner of publication (and non-publication) and style of work. The style expressed the man in Newton's work as much as in Milton's: the simple perfection of his experimental style, the lofty archaism of both language and mathematics in the *Principia*, the stiff, impersonal, official style he imposed on the Royal Society, and through that on all British science. The *Principia is* Newton, as much as *Paradise Lost is* Milton; and both, of course, are 17th-century England. That is what courses in science-and-history should convey.

Turning to the other general category, science-and-literature, we should have courses in science fiction, which should sharpen the students' sense of genre and sub-genre, the way a formal idea first rendered by one writer can be fruitfully developed by another, and again by another, up to a certain point, and after that ceases to be fertile. Just because of the numbers and the easy readability of science fiction novels, it is easier to study them as a form than it is literary novels; and the course can keep putting that other form into perspective by comparison and contrast — in terms of date, nationality, politics, readership, as well as characterization, style, incident, etc. And again because of the multiplicity and the facility of the examples, it is easier to make certain kinds of qualitative distinction than it is with literary novels. Certain kinds of superiority in narrative skill, in structure, and in style, occur spontaneously to the reader, and the sense of them can be sharpened to the point where it becomes useful in reading the other kind of fiction.

Then I think there should be courses in the scientific novel. One can contrast the English tradition, seen in H. G. Wells' *Tono-Bungay* and C.P. Snow's *The Search*, for instance, with the traditions of France, Germany, Russia, America. The comparative rarity of good quality scientific novels in English is itself an important subject of inquiry, while the changes in form and style, from country to country and from decade to decade, are in some ways easier to recognize and discuss in work of the second quality. The differences between *The Search* and McMahon's *Principles of American Nuclear Chemistry*, for instance, illustrate many changes in the character of the ambitious middlebrow novel between England in the 1930s and America in the 1970s. In the second, the character Mary Ann is symbolically related to the freedom of the South-Western desert where the atomic bomb tests take place, and to the afflatus of the scientists assembled there and for once given unlimited funds and official sponsorship. This is a use of character which reminds one of Lara in *Dr. Zhivago* and Tamara in *Speak Memory*. Both those girls are symbolically related to the promised freedom of the Russian revolution, by Pasternak in the one case and Nabokov in the other. In the scientific novel, the girl and the larger cultural meaning relate to the central character in the same way and amplify each other, just as in the two Russian novels. Such symbolic characterization is unimaginable in *The Search*, as is the dramatically wounded character of the central protagonist. The contrast between the two kinds of scientific novel is a vivid example of the way the world's literary imagination has changed over the years. By combining that with the scientific change between the physics of *The Search* and the physics of *Principles of American Nuclear Chemistry*, one can show the movement of cultural history itself.

One could also give a good course in intellectual autobiographies. Amongst the scientists' work, G.A. Hardy's *A Mathematician's Apology* and Norbert Wiener's *Ex-Prodigy* and *I Am a Mathematician* would be exemplary texts. One wants books like these, which make some attempt to explain their mathematics in itself, as well as in its effects upon their lives. Then these might be set in contrast with a historian's autobiography, Gibbon's or Collingwood's, and a poet's or novelist's, like Yeats, James, or Nabokov. The latter should of course be taught as the way the writer acquired his imagination, his vision, so that the bravura passages in, for example, Nabokov, can be set in contrast with the mathematical passages in Hardy and Wiener.

Then we could have a good course in the popularization of science, which would read a number of examples of this genre, and would try to establish some criteria for admiring some and not others. It is striking how lacking such criteria are, how unaware editors, for instance, are of what makes satisfying popularization. It is unsatisfying, for instance, to have loud jokes made throughout a difficult piece of exposition. It is unsatisfying to have references to juke-boxes and the twist offered us as a consolation for our uneasiness with a subject like photosynthesis. It is unsatisfying to have commonplaces of history (the Armada, let's say) described to us in picture language. It is unsatisfying to have some scientist's problem compared for our benefit with his cat's effort to get some fish out of a covered dish. In short, the average popularizer addresses himself to "the child in all of us." For instance, Isaac Asimov has done some good work in science-for-the-layman, but there are times when he makes one feel curled up in one's cradle being taught to count by a twinkly man who is pinching one's little toes one by one. To establish the superiority of popularization that addresses itself to adults, and that makes science again a part of the intellectual and imaginative adventure of the Western mind, would be both an excursion into literary values for the science student, and a useful contribution to the culture at large.

The most ambitious course I would like to attempt would relate, for instance, Einstein's work on the General Theory of Relativity to D.H. Lawrence's work on *Women in Love*. The two bodies of work were contemporary, and both were in some sense revolutionary, but the two senses were very different. Such a course would describe and explain in turn each man's achievement, in its setting of previous literary or scientific history, but also in its setting of subsequent developments. It would also describe the two men's personalities and careers, with primary stress on the intellectual aspects in each case, but quite freely exploring beyond that to present a complete picture. It could also describe the European scene that divided Einstein's Switzerland from Lawrence's Cornwall in 1916; the slaughter on the Somme. There is no need for any striking moral to emerge from the comparison. All that is needed is that each should be fully presented in its historical setting, as an achievement of the Western intellectual tradition.

All thirteen of these courses are ways of reintegrating science into the humanistic imagination. Orthodox history and philosophy of science do that in their own way, but they often seem to be inaccessibly withdrawn within their own technicalities. What we need is a non-technical and lively atmosphere of ideas about science-in-culture — like the atmosphere of the Wells, Hogben, Eddington books, but without their ideology — and courses of this kind seem to me a good way to achieve it.

Colloquium: The Role of the Humanities at M.I.T.

The Function of the Humanities at M.I.T.

Richard M. Douglas
Professor of History

M.I.T. was founded on an audacity — on the belief of the founder that if science and engineering were to acquire a strong and significant place in American education, the study of these fields should be placed at the center of a new kind of university curriculum rather than at the side of it. William B. Rogers thus repudiated the prevailing doctrine at Harvard that "professional training should be added to rather than combined with the traditional liberal arts curriculum." It was Rogers' determination by 1861 to create an autonomous and free-standing position for applied science which led to the establishment of a new phenomenon in higher education in the United States. This curriculum was based on five professional courses of study: Mechanical Engineering, Civil and Topographical Engineering, Practical Chemistry, Geology and Mining, Building and Architecture. The sixth option was known as General Science and Literature, intended to "furnish such a general education . . . as shall form a fitting preparation for any of the departments of active life."

At the beginning, it was simply assumed that the study of literature at Boston Tech was to be pursued for its own sake. Literature and modern languages constituted the humanities. They belonged to two admittedly separate cultures, one then known as "classical," the other "utilitarian." Only later was the derogation "service department " applied to the humanities. My own position is that Rogers was basically right. He was right in his apparent assumption that the liberal arts at M.I.T. be accorded the same essential autonomy which he labored so long to achieve for engineering. It is not that humanistic study is to be regarded as antithetical to science and engineering, or as something "counterbalancing" or "counteracting" the pursuit of professional education and training. It is instead a matter of saying that courses called "English for Engineers" are invariably an intellectual scandal. And "Economics for Engineers" is just as often a debasement or a denatured act of accommodation. If humanistic teaching and scholarship are to make a difference in a university of science and engineering, they must be chartered and sustained with the same basic respect for their "own-sake elements" as economics or linguistics and by the same understanding that the achievement of mastery and competence in humanistic study requires the same kind of discipline and difficulty as in the case of science. I am merely making the same fundamental proposition for humanities which is claimed in the Report of the M.I.T. Commission on Education in behalf of science: "all of us at M.I.T. believe deeply in the value of science for its own sake, as the expression of an age-old desire to know." (p.47)

This is not to deny at all that a good department of humanities in a university of science should enjoy any warrant to disregard the culture of science in teaching courses, for example, on the history of thought. Far from it. Done right, such courses should — indeed must — include a steady and active participation by scientists. (Our own efforts in this direction have been timid and feeble to date, but help appears at last to be on the way.) The eminence of DeSantillana in the history of science, however, was derived as much from the capaciousness of his literary culture as from his scientific understanding. The point is that those who urge the integration of science and literature must be chastened by the rarity of DeSantillana's example, the example of a man as much at ease with a poet as with a physicist.

Nor do I mean at all to deny that institutions like M.I.T. have a special obligation to the history of science and the history of technology at the level of research as well as teaching. What is so obvious, however, about successful efforts at interdisciplinary courses is the spontaneity of their origins and the specificity of their subject matter. Such courses do not arise from injunctions to bring scientists and humanists together in a seminar on "Knowledge and Values" but rather from shared curiosity about the Crab Nebula — or the use of nuclear power stations on Lake Cayuga or the Hudson River.

I would be reluctant to offer a definition of "*the* proper function for the arts and humanities at M.I.T." Seen by my colleagues, the purposes of humanities here range all the way from the pursuit of mastery over a discrete discipline to a missionary zeal about pushing "the king's culture on angry subjects." I do know that it is far easier to understand what is meant by mastery over a subject — to agree on the implications of that term — than it is to be sure what Charles Frankel meant, in talking to the M.I.T. Commission, when he observed that ". . . liberal education has the potential to give a person defenses against being stupefied by his experience in later life." Nonetheless, I *feel* intuitively that Frankel is right. I believe that the study of the liberal arts, when it works right, is like a successful psychoanalysis: one is indeed less likely to be stupefied by his own experience in either case, whether that experience issues from living-through-history-and-time, or from living-among-other-people.

Everyone's life today, whether he likes it or not, is a version of what the Romans called the *vita activa* as distinguished from the contemplative life. The controversies which bedevil all humanities departments these days involve the question of how indeed to "furnish such a general education . . . as shall form a fitting preparation for any of the departments of active life." How indeed to maximize the potential of humanistic education within the constraints of very limited time, and how in practice best to serve the several purposes of liberal studies in an environment of professional training? How in fact to combine, as Rogers put it in an earlier idiom, the "classical" with the "utilitarian" curriculum?

To speak of a single purpose behind the teaching of humanities in the environment of Cal Tech, Case, Carnegie-Mellon, or M.I.T. would be altogether simplistic. Moreover, because the term "humanities" is, in practice, a cable address for humanities and social science, the student's elective options range all the way from the "Bible" and the "Epic" to "Prices and Production" or "Neural Mechanisms of Learning." Even within the disciplines of the humanities alone there are vast and obvious differences between a fourth-level course in music theory and a seminar on the Cold War.

The term "humanities" refers not only to a group of identifiable disciplines but also to those studies whose primary concern is to enlarge our understanding of what it means to be human, and to widen our awareness of what humane values are both in the abstract and under particular circumstances of time and place. The word inevitably contains a relative meaning, which has shifted frequently from its invention in Antiquity to its rediscovery in the Renaissance, and no less out of the intellectual convulsions of our own time. Whether in the case of Cicero, Erasmus or Malraux, the humanities have always served a dual function in the hands of its more influential spokesmen: studied on the one hand for themselves, and then in relation to something else — where something else has often been expressed with the voice of cultural criticism, addressed to changing experience and historical circumstance.

The center of the problem in virtually all liberal arts colleges and universities, and no less at M.I.T., seems to be the curriculum of the first two years. For a variety of reasons, one listens in vain for voices proclaiming master remedies with which to clear away the accumulation of disparate courses and programs whose apparent purposes can be inconsistent, incompatible or contradictory. The so-called "crisis in the humanities" has been brought on in part by the proliferation of subject matter in courses during the '60's. (How many colleges still settle for Western Civilization and English Composition as the two major staples of the freshman year?) But it has come about no less from an enormous fragmentation within the intellectual tradition, or — to put it more dramatically — from conflict, probably unprecedented in this century, within the academic community itself over the right definition of humanistic teaching, scholarship and study.

Perhaps the curriculum is condemned to a disordered kind of pluralism for a while. During the time of expansion in American universities, there was a visible tendency to add new programs without changing or abandoning the old ones. But today, as Fred Hechinger recently put it, "in the absence of exciting new goals, confidence has evaporated." As he also points out, many of the new pressures are negative. These are harsh truths, but they must also be acknowledged by anyone who wants to join realistically in a search for solutions. The condition of the humanities curricula in universities of science represents an intensification of the problem in liberal arts colleges, where conflicts so often are resolved by allocating time and space in the curriculum to all parties to the debate. There, the student, after all, has four years in which to work out his own salvation. In engineering schools, the student and the humanities faculty have much less than the equivalent of one.

On Humanities Requirements for
Students in Technical Curricula

The question of requirements can be described very quickly. First, should they exist at all? And if so, how narrowly? The American Society for Engineering Education provided the answer to the first, in general terms, which is observed in all accredited engineering schools. The so-called Hammond Report of 1940 and 1944 (*Journal of Engineering Education*, March 1940 and May 1944) recommended the required equivalent of at least one-half year of the usual four-year undergraduate curriculum in humanities and social science, and assumed a condition of parity in importance between these two subject areas.

How narrowly or how permissively? Practice has varied widely in humanities departments from complete prescription to multiple options at M.I.T. (now eight full-year courses in the first year and nine one-semester courses in the second year of the "core curriculum"). The requirement for juniors and seniors consists of three courses in one of twelve fields of humanities and social science, and one course for distribution in a second field. What is at issue is not so much the legitimacy of the minimum requirement as the implementation of it: whether to maintain, enlarge or abolish the restricted options in "general education" for freshmen; and whether or not to permit and encourage the selection of a single discipline for "concentration" beginning at the outset of a student's career. Is the proper model to be found in the old Reed or in the new Brown? In either case, the number of available semester-courses remains at eight.

Those who defend "general education" in its original conception do so on the argument that interdisciplinary study is necessary at M.I.T. because of its special character and because of the Institute's strong commitment to professional and highly specialized training. Given the brevity of time for humanities and social science and the lack of leisure for exploration at the beginning, proponents of a structured body of requirements also express concern that the "classical" or humanistic option might otherwise be repudiated by the kind of student who wants his entire program, even in the non-science areas, to be "utilitarian" or instrumental.

Those on the other side question the educational gain of any kind of coercion, however well-intentioned, and observe the resentment of bright students who are compelled to do one thing while yearning to do another. Here again we share the dilemmas faced in the colleges. Given the fact of disagreement about common purposes, with the disintegration of shared goals and indeed even of a shared way of talking about goals; faced with the proliferation of programs and the proliferation of viewpoints or ideologies within single fields; and confronted at the same time by a more sophisticated species of undergraduate — bright, restless, often uncertain about his own purposes as a freshman or sophomore, the temptation can be strong here as elsewhere to let the student invent his own commands and devise his own program of study.

The major issue over requirements at M.I.T. is whether to curtail or eliminate the curriculum of the first two years as it now stands, allowing the student to begin a program of concentration in a single discipline from the outset, with the stipulation that those whose field is one of the humanities select two additional courses in a field of the social sciences, and *vice versa.*

The present program requires one full-year course in general education for freshmen, followed by the requirement that sophomores select a one-term course from a list which includes anthropology, history, and political science, and from another list which includes literature, music, and philosophy.

In principle I support the latter alternative on the ground that it provides wider perspectives for choices to follow when a student selects a discipline or field of interest for concentration. It strengthens rather than weakens the potential for intellectual collaboration in the development of new courses. And it expands the idea of teaching to something more inclusive than the "introductory course," serving parochial departmental purposes.

Engineers
and Humanists

David J. Rose
Professor of Nuclear Engineering

Good science and good engineering are arts in themselves, and that comment is made more than in passing. A gulf of difference lies between plodding pedestrian articles in *Physical Review* and the insight that leads to understanding what a pulsar is; or between bridges on the Merritt Parkway (now mercifully overgrown with ivy) and the bridges painted by Corot or Monet. Such matters have to do with developing a sense of taste as well as developing an I. Q.; so many more directions and possibilities being ahead than we could possibly follow, we require before doing the work to have truly an artistic sense of what should be done. The good scientist and good engineer guide their work with sensibilities and sensitivities that are basically artistic. Thus a proper education at M.I.T. must include the arts and humanities.

Many argue on the other hand that the exquisite taste of which I speak does not involve the arts or humanities at all, but rather the development of consummate skills. Hence the bookish scientific recluse who lives only to advance knowledge in his narrow area does his job quite well. That point of view can be defended tolerably well for pure science (although I think it places the practitioner at a severe disadvantage), because pure science *can* to a considerable extent be divorced from society.

But not so for engineering. Here I define the proper role of engineering as the use of science and technology to satisfy the perceived needs of society. Thus the engineer does not just build the bridge: he responds to a need to get from here to there, and maybe that particular bridge wouldn't be part of the best path. Even more, asking why you wanted to go on the trip in the first place can be part of the engineer's legitimate business. The engineering handbooks don't often teach those things.

What I come to in this response is my desire to illuminate the connectedness of things; of engineering and scientific tasks with societal goals and societal activities — and how can one do that without an understanding of what our civilization is like? Take history, for example; I read it as a favorite avocation, and learn much about former times that causes me to reflect upon what we do today. The classic empires around the Mediterranean came to an energy crisis through exhausting their supply of the only fuel they knew how to use (wood), which was a contributory reason for technological superiority moving north to still-forested lands. I read that a larger fraction of English houses had central heating in 400 A.D. than did those in 1930. And that a larger fraction of English college-age students attended college in 1630 than did in 1930. Neither mere passage of time, nor change, nor more of anything, necessarily mean progress — however defined — and some understanding of history helps one to develop perspective in these matters.

Besides history, one can identify many other fields of study: witness the peculiar role of architecture, as it incorporates engineering design, utility, and aesthetics. Or consider the plight of the Atomic Energy Commission, as it now repairs its intellectual position about disposing of radioactive wastes — a difficulty that it got into by imagining the problem to be purely technical, and failing to see that the chiefest considerations were societal ones. Or consider present environmental arguments — many of them really turn on different time scales of concern in different societal sectors; one cannot understand such matters without a quite humanistic approach.

The humanities must be woven into the very fabric of the technological material. One cannot put controlled fusion research in proper perspective without knowing why we need energy in the first place — and perhaps we don't need *that* much energy. The atomic theory takes on new meaning and stimulates new interest if one recognizes its continuous development since classic Greek times.

"Le Coq", 43" x 18¾" (Woodcu

Then too, this country suffers from a surfeit of incredibly unreadable, sender-oriented scientific and technological reports, almost none of which will have any effect whatsoever. One cannot get a message across except by being receiver-oriented, and development of literacy, let alone eloquence, is a most necessary and central business.

This leads me to suggest that in the applied sciences and engineering particularly, there is a need for new courses that teach the differences between means and ends, between past conditions and present ones, between doing technological design and communicating balance among technological options. One factor that bears on this is the lack of available time in undergraduate curricula. I think that professional education moves more and more into the graduate years, as time goes by. Thus the undergraduate years become available for a more general preparation — pre-engineering, say — wherein socio-technological assessment, historical perspective and literary skill naturally get taught. The humanities are much more than mere dispensable items. "How 'liberal' can undergraduate education become?" you ask, presumably as a consequence of pouring in more humanities. If by liberal you mean "broader than technology" and "societally conscious," then for an answer, I do not know how far these matters should go. I need not worry just now, because operationally one can take several steps along the new path while still being certain of not having gone too far. From there, we can see ahead a little better what to do.

I would like to see us make the relationship of technical disciplines to human and social values a naturally conscious one, to wit: just as an engineering student will learn calculus because he sees it fitting into a useful pattern, so he or she will see relations between the technical disciplines and humanistic (or social) values, if these matters are naturally included. Here are two examples. The first is simple; I recommend Dorothy Nelkin's book on the Cayuga Lake Controversy, as an example of the importance of very non-technical issues, in the siting of electric power plants. In there, we find development versus conservation, lack of communication between New York State Electricity and Gas Company and the Committee to Save Lake Cayuga; both sides were partly right, partly wrong; and the attitudes of both sides caused them to talk right past each other. (*Nuclear Power and Its Critics: The Cayuga Lake Controversy*, Cornell University Press, 1971)

The other case is my own at M.I.T., attempting to teach technology assessment (graduate students mainly, but these thoughts apply to undergraduates also). About one-third of this course is broad technology assessment, one-third relates to analysis of the shortcoming of institutions (e.g. universities, national laboratories, etc.) to deal with large national problems, and one-third relates to social, historical, and cultural aspects of both the problems and the institutions. The students seem to benefit, and so do I.

We have students in our Nuclear Engineering Department who have switched from straight economics of nuclear power, or plasma physics, or reactor engineering, to thesis topics on:

how one quantifies the societal value of land, to attempt a more rational analysis of present coal strip-mining;

cities as energy-users, hence (in part) energy related polluters, hence (in part) hostile environments; and

international implications and difficulties likely to arise because of developing disparate policies on radioactive waste disposal, and failing to connect them with possible future societies (that must accept what we impose on them).

Perhaps these topics seem not to go very far toward total integration of the humanities into what has been a technological education; but from where I sit, such things look pretty good.

M.I.T. and other places now talk openly of establishing these intellectual bridges. Of course, what we would be seeing would be not the making of something new, but the restoration of something that had existed at other times and places. As you surely know, the narrow specialization in the numerical and material aspects of science and technology leads to much and powerful methodology — but only to develop very narrow solutions. These points are being recognized not only in the School of Engineering at M.I.T., but I have even had the pleasure of seeing such ideas accepted (often grudgingly) at National Laboratories. At M.I.T., we have taught about the history of the atomic theory, about energy and society, and about the interaction between national purpose and technology, all of this by professors of nuclear engineering.

Humanists have, unfortunately, not always been in the van of these movements, and it has often fallen to the engineers or scientists to try bridging the gap. And so it seems at M.I.T. presently. If that stolidity of the humanists be real, all it means is that engineers and scientists will have the better education for their trouble. But I think that the humanists will actually find it interesting to associate their skills with those of technologists; intellectual content exists at the synaptic points between the various threads of thought. Just as we technologically-inclined people benefit from associations with the humanists, so do they, *mutatis mutandis.*

A Bridge:
The History
of
Technology

Brooke Hindle
Director
National Museum of History and Technology
Smithsonian Institution

The role of the humanities in undergraduate education at M.I.T. deeply concerns many of the faculty and the administration. It remains an unresolved question despite several approaches, some of which have been pioneering in character. Today, a new urgency surrounds the question because the present complex of crises engulfing our civilization seems peculiarly related to man and his science and engineering and, inescapably, to this leading technical university.

Some who have confronted the need for a humanistically oriented education at M.I.T. have seen the institution as a primary cause of our problems; as hardened upon a course which celebrates power as its objective and is often blind to human values. Such criticisms are not merely wrong; they obscure the real difficulties. Strong men differ starkly on many of the critical issues, but the institutional problem is not at all one of willful blindness. The problem is that no part of the solution is obvious or available. M.I.T. has an opportunity to find a path which could be as important to civilization as any of its past achievements — but the path remains to be discovered.

During the academic year 1971-72, I served as Killian Visiting Professor at M.I.T. — on leave from New York University. My field is the history of technology and most of my research and writing has been in early American science and technology. My outlook is that of a general historian, convinced that one of the most critically deficient areas of our historical understanding is the history of technology and anxious to integrate technology into the fabric of our historical understanding.

This focus is upon the reverse side of M.I.T.'s problem. I have been more concerned in seeking ways to introduce technology into our historical synthesis than in introducing history and the humanities into science and engineering education. The two, however, are not merely related, they represent complementary aspects of the same question. In each case, to use a familiar analogy, the need is for a bridge.

I believe, and have argued, that the history of science and technology constitutes one such bridge — a two-way bridge that can be used to connect the worlds of science and technology with the humanistic disciplines. There are also other bridges but I feel best qualified to comment upon this one.

A View of Humanities at M.I.T.

The humanities have had a unique and varied experience at M.I.T. Here, humanities include everything outside of science and engineering — the social sciences as well as those studies usually identified as humanities. M.I.T. led in expanding the attention given humanities in a science-engineering curriculum, but the great institutional emphasis has remained on the sciences and science-based engineering. The early "Department of English and History" was a service operation expected to dispense cultural gleanings but not to bulk large in the student's effort or in the Institute's intellectual structure. Later, in a significant effort to strengthen humanities, a fine group of young Ph.D.'s worked on a teaching program designed to fill M.I.T.'s particular needs — and continuing efforts have followed. That group looked forward to individual research careers and most of the men went on to signfiicant achievement, many to other institutions. Indeed, the honor roll of humanities faculty alumni is outstanding. Among those who have attained professional recognition at M.I.T. are members of the teaching department of the 1930's who developed their research to a point of leadership — especially in the history of technology where the stimulus of the environment is so great.

Yet for all this, a feeling persists in the Institute that the humanities department is not comparable in distinction to electrical engineering or mathematics, and, more to the point, that the insights of the humanities have not sufficiently permeated either undergraduate education or the general intellectual atmosphere.

Quality vs. Relevance

The problem may be viewed as twofold: first, the development of a level of excellence and professional leadership among the humanities faculty comparable to other sectors of the Institute and second, the design of curricular programs and requirements which will relate the humanities to the primary concerns and values of the undergraduates. The attainment of a high level of quality seems imperative to permit the humanities to speak effectively in the achievement-oriented atmosphere of M.I.T. Only then can there be hope that the values of the humanities might pervade the faculty and student scene to the extent that they do in a good, broadly developed university. At the same time. it is essential that the undergraduate feel that his humanities courses are more than an extraneous potpourri; they must seem relevant to his own life and to his major interests.

The perceived need for quality has led to remarkable achievement in developing areas of excellence outside the realms of science and engineering. Strong support and astute planning, for example, have brought to M.I.T. a top-level economics department. Similarly, linguistics and political science have been brought to leadership. So far, the most notable achievements of this sort have been in the social sciences or in science-based fields of a different character and with different values from the traditional humanities. Recently, however, philosophy has been established as a separate department with a graduate program and aspirations to the attainment of excellence.

Within the humanities it is imperative that some areas attain the highest quality. Faculty members must be leaders of their profession and able to participate equally in the intellectual interchanges which are so much a part of the Cambridge environment. Today, the M.I.T. community is remarkably full of discourse on all sorts of questions and more than ready to receive humanistic inputs. Indeed, my own experience suggests that, perhaps sensitive to the criticism of over-specialization and limited breadth, scientists and engineers here are *more* ready to discuss large questions beyond the bounds of their professional competence than most historians. Receptive as elements of the community are to the humanities, they will be reached only by those respected for their own competence or ideas. Students, too, are keyed to the respect for quality and professional leadership. Upon the base of several areas of excellence, the humanities might at this moment of time expect to have considerable influence both directly through course work and indirectly by altering the general intellectual atmosphere.

The attainment of such quality is possible at M.I.T. only by certain routes and it carries certain limitations and difficulties. Concentration upon a few fields seems imperative for it is not realistic to think of developing as many areas of excellence in the humanities as some general universities may boast. Moreover, wherever excellence is sought, it has to be accepted in terms of that field's directions of motion. For example, an excellent economics department has to be excellent in terms of those aspects of economics that are important to leading professional economists. It could not reach excellence by offering a good sequence in "economics for engineers." Even more antithetical to quality would be an entire humanities program based upon selections of "humanities for scientists and engineers." Quality has its own rewards but it does close out some possibilities.

At M.I.T., there is no alternative to the attainment of excellence in selected areas of the humanities. There is no other way to encourage increased influence of humanistic scholars upon the Institute community; there is no other route by which humanities can become a central part of the undergraduate's experience.

Relevance

Excellence, however, is not enough. More is being asked of humanities at M.I.T. than is demanded in a general university or than is asked here of geology or nuclear engineering. In the science and engineering departments, a kind of internal autonomy is possible because their role is largely limited to their own disciplines and to those closely related fields with which they are naturally connected. They feed and draw upon other departments but they are not asked to impress their values upon all undergraduates or to permeate the entire Institute. On the other hand, the humanities are expected to offer more than professional excellence. They are expected to provide human values and insights to the undergraduates and to leaven the entire Institute.

Outside M.I.T. American collegiate education is well along the road to giving up all undergraduate requirements, aside from major or concentration requirements and unavoidable prerequisite demands. Some institutions have actually reached this point. In few others is the attempt still made to design and define the whole program. Related, coordinated sequences in both the humanities and the sciences have become rare. This movement is rationalized on the grounds that no one can any longer presume to define the outlines of a liberal education and that each student should have a program tailored to his own needs.

At M.I.T. a sense of responsibility for the entire undergraduate experience persists, related no doubt to the unavoidable necessity of demanding specific sequential and interrelated course patterns to provide scientific or engineering competence. Now, there is much feeling that something more than specific professional competence is needed, however creative or brilliant. It is presumed (correctly, I believe) that graduates may lack some of the humanistic insights and values which graduates of good liberal arts colleges appear more often to have. It is reasonably presumed that the humanities component might be increased to satisfactory proportion by better curricular design.

Unfortunately, experience suggests that offering excellent courses by excellent departments may not be enough. This reaches some undergraduates but for many remains outside their own genuine interests. Tailoring courses and sequences to the assumed needs of M.I.T. students can also miss the objective when it ventures too far from the major directions of development within the humanities disciplines involved. Excellence in the humanities is imperative and so is a relationship to the student's own interests and pursuits. Of course, the wider the spectrum of humanities options offered, the more students can be reached effectively. However, there is a limit to the diversity of humanities areas M.I.T. can bring to a point of excellence.

The Bridges

One means of combining planned excellence in the humanities with relevance or relationship to the interests of students has been discovered at some points but not systematically studied or developed. This is concentration upon those bridges between the sciences and the humanities which identify themselves in the M.I.T. community. Certain areas of the humanities seem to have an intrinsic appeal and support, among them the history of technology, science policy, and the history of art and architecture. Similarly, though not developed here, the philosophy and sociology of science might serve as natural bridges. Each study of this sort is a sub-discipline which could be developed to the point of excellence and given a large role in the curriculum without attempting to develop the full disciplines in any similar way. For example, the history of technology might be brought to the highest level of excellence while history as a whole remained representative and less concentrated in its development.

Within the School of Architecture, activity in the history of architecture, and recently in the history of art, has been significant. Art finds a curious responsiveness in the scientist and the engineer — perhaps because its non-verbal character touches the non-verbal components of technology and science. At any rate, the history of art provides one recognized bridge between science and the humanities which has great potential for development at M.I.T.

Another such bridge is the history of science and technology. Before they emerged as distinguishable disciplines in their own right, both the history of science and the history of technology had evoked much response at M.I.T. Historians as well as scientists have made contributions here and the drift toward this interest has never ceased. In view of the record and of the clearly stimulating environment, it is a matter of frequent professional comment that M.I.T. has not supported either effort to the level of excellence it could easily attain. Growing so naturally out of the M.I.T. environment, the history of science and technology provides a bridge which relates the attitudes and understandings of an area of humanities to the central interests of the science and engineering student. It would not be necessary to raise all components of history to the same level of excellence and no selection or distortion would be required to relate this study to student professional interests. Indeed, the history of science and technology provides exactly the perspective that is so conspicuously needed in the community and in the curriculum today. It offers the dimension of depth in time for the understanding of modern science and technology. Both are young fields so M.I.T. has the greater opportunity to become a major contributor in this development.

Some Thoughts about the Humanities at M.I.T.

Judith Wechsler
Assistant Professor of the History of Art

It would be easy to caricature science as "technical education" and humanities either as cultural social work or as "dispensable luxury items;" to some extent the questions posed for our response in this colloquium did so. It would be more constructive to bear in mind that science is a qualitative experience as well as that humanities have disciplines with their own rigors. An appropriate question might be how the sciences and humanities can address themselves cooperatively to the larger issues which are at stake. To this end the sciences and humanities should re-evaluate their roles vis-a-vis each other within educational institutions.

Questions regarding the quality of our contribution and understanding in regard to culture should not be approached abstractly. Without intensive study of the particular scientific concept, experiment, or work of art in which broader humanistic issues are reflected, the questions remain so general that they can only receive vague answers and thus will not change the nature of our outlook within our own fields. At tne same time the universal character of scientific inquiry is often minimized in specialized study and research. A similar situation prevails in the modernist approach to scholarship and criticism in art and literature.

The humanities should not blindly emulate scientific approaches, just because they appear more substantial and respectable in a community like M.I.T. Nor should science study become generalized so that research is rendered inconsequential. One must know to what purpose an interdisciplinary approach is being used or encouraged. Science has positively affected the humanities by its example of careful examination of evidence. Humanities, social science, and arts can shed light on different kinds of inference and can provide ways of understanding historical contingencies and continuities.

Curiosity, human understanding, and responsibility are integrally a part of scientific inquiry and research. In *Physics and Beyond* (1971), Werner Heisenberg records that the discoveries of relativity theory, quantum mechanics and elementary particles were affected by philosophical, religious and aesthetic perspectives. The problem is how to develop an awareness and knowledge of the role of science in culture and civilization without sacrificing a high degree of technical education. One way of doing this might be to introduce, in addition to the required courses in an area of specialization, a parallel course to study the historical foundation of the field, its intellectual history, and changes in its cognitive models. Studying the assumptions, approaches, and directions of a field from political, economic, cultural, and social as well as scientific points of view can minimize the chances that a student will regard his discipline in a purely technical and instrumental way.

It would be unrealistic to expect students to be able to carry out this highly complex integration of disciplines on their own. A semi-conscious unarticulated sense would be insufficient. When awareness and knowledge are generated in a concrete way within the area of specialization a sense of human values has more likelihood of developing logically within one's work.

A basic course like the proposed one is by no means meant to take the place of most present humanities courses at M.I.T. The proposal is offered as an example of how the humanities and science faculties could work cooperatively to their own mutual benefit and that of their students. The concerns of the humanities and sciences are not diametrically opposed. Historical and aesthetic knowledge and understanding which is developed in the humanities should become an integral part of scientific education. The humanities should not be viewed as "dispensable luxury items" any more than a science program is dispensable in a liberal arts college. With the exclusion of either we propagate a distorted view of our culture.

Regarding experiments in humanities instruction, in the fall of 1972, I organized a course, "Topics in Art, Science, and Technology," which brought together scientists, historians of science, and art historians to examine ways in which one could regard aesthetics of concepts, processes, elements, materials, and products in science and the reflections of scientific concerns in art. Each week's lecture was followed by discussion. Sixty students registered for the course and up to 300 auditors attended. The students were asked to formulate and develop a paper or project concerning an aesthetic perspective in their own field of study; they were to find advisers within their own departments. Almost without exception the professors contacted were supportive of the venture. The opportunity for students to view their subject from a different perspective was enthusiastically received; the range of topics for the term project was wide and imaginative and the level of work extraordinarily good. The response has been encouraging and the course will be repeated annually, with varying participating lecturers from different areas of science. A list of student topics may convey the character of the course.

Selected Student Projects:

The Hele Shaw flow principle in fluid mechanics was demonstrated in a model which illustrated complex and often geometrically ordered patterns of flow fields.

The ionization of different gases made visible through a constructed model accompanied by numerous photographs documenting the forms and colors emitted from the gases and mixtures.

Natural selection and the glass sculpture of algae: an analysis of why algae take on a certain shape and exhibit patterns of ornamentation.

The development of a new construction material and the problem of how to design a mix, and predict the final properties, structural and aesthetic, that the transformation process will cause.

Visual representation of the structure of various bacteriophages and animal viruses, with the aid of an electron microscope.

The aesthetics of making microscopic slides — different staining techniques of biological tissues.

The heart, levels of aesthetic value in form and function: teleology as a factor in the aesthetic quality.

Observations outside the visible range in astrophysics and the photographic means of a translation to a digital form.

The aesthetics of integrated circuits.

Cartography: the art and science of map making.

The relationship of the study of anatomy and art.

On the origin and development of the scale in Eastern and Western music and the mathematical analysis of harmonic frequencies.

The Chinese city in history, the evolution of its form in relation to societal and technological changes.

The evolution of the technology and aesthetic of watches.

Frieze pattern symmetries: an index of civilization.

The development of perspective theory in painting.

"Birdbox", 29" x 38" (Woodcut).

The Integration of Knowledge: Notes on Curriculum Change at M.I.T.

Christopher Schaefer
Assistant Professor of Political Science

One

The Commission on M.I.T. Education stated that the humanities and sciences were not well integrated at the Institute. This understatement does not do justice to the isolation of the humanities within an institution dominated by a scientific and technical ethic. From the time of their addition to the M.I.T. curriculum, the humanities were perceived as a way to round off the rough edges of a scientific education. The primary function of the new Humanities Department was to provide the M.I.T. undergraduate with a veneer of Aristotle, Raphael, and Keats.

This situation has not basically changed. The general attitude toward the humanities at M.I.T. continues to be one of condescension, based on the belief that the humanities do not say useful or important things about man and the world because they are not scientific. This attitude is partly due to the nature of the Institute; its technical orientation, finances and admissions policies, and partly the result of the cultural values esteemed by our scientific civilization. It also reflects a persistent lack of courage among humanists generally. All too often they have responded to the implicit challenge of science and technology by either glorifying past cultural achievements or by attempting to give their field the aura of scientific respectability. In the first case, they condemn themselves to irrelevance by not speaking to the present, while in the second, they abandon the meaning of their subject matter. Counting the number of times "frolic" appears in Bocaccio's *Decameron Tales* may assert one's ability to count, but it surely detracts from the meaning and joy of the stories. The fact that scientists such as Rene Dubos and closer to home, computer specialists such as Joseph Weizenbaum, feel compelled to speak in the name of humanism suggests that the voices of traditional humanism are barely audible.[2]

Despite the uncertain position of the humanities at M.I.T. and within our culture, I believe that they must become central to undergraduate educations if we are to overcome our social, environmental and human crises. However, to achieve an integration of the humanities and the sciences will be extremely difficult. Even if a concerted desire to bring about such a change existed, the necessary sustaining vision of human knowledge is lacking. As long as our concept of knowledge remains so intricately connected with sense-based scientific knowledge, the humanities will be regarded as interesting but secondary fields of human endeavor. Because of this I can only envisage a new integration of science and the humanities taking place if a different and shared image of knowledge emerges. The description of such an image, besides being hazardous, is clearly beyond the scope of this essay. Nevertheless, I wish to outline some preliminary aspects of such an image, for without it educational reform is haphazard tinkering with tradition.

Two

The image of knowledge which I wish to draw attention to is critical, reflexive, affirmative and moral. By critical reflexion I mean knowledge grounded in a foundational philosophy which seeks to demythologize knowledge and belief by reflexively raising to consciousness the assumptions, the habits of perception and thought by which we experience the world. I think that the path toward such a radical-making-consciousness was indicated by Edmund Husserl, the founder of phenomenology, and in a somewhat different way by the Austrian philosopher and educator Rudolf Steiner.[3]

One of the important efforts of such a foundational philosophy must be to question the idol of science as a world conception. By this I am not suggesting a rejection of natural science or of the scientific method, but rather a clarification of its assumptions, ideology, history and impact. A number of illustrations come to mind. Science assumes a separation between the scientist and the object he is studying. It also extrapolates from sense data, not basically doubting its validity. Yet quantum mechanics has revealed the observer effect in studying the basic units of matter.[4] We are also increasingly aware of the degree to which science gives answers to the questions it asks, and are beginning to appreciate the selectivity of our sense organs. This awareness should foster the recognition that, "No mosaic constructed from messages thus highly selected can show more than reality's skeleton."[5]

If the systematic pursuit of such questions and the exploration of the differences between the scientific world picture and our experience of the world is not a sufficient antidote to the hubris of a scientific ideology, then perhaps the historical record can be helpful. In the course of human evolution the image and experience of reality has fundamentally changed. Why is ours the correct and final perspective?

By raising to explicitness the assumptions and practices of science, the incompleteness and selectivity of the scientific world picture can be made manifest and its unwarranted extensions combated. Perhaps then the warning against the consequences of idolatry contained in the 135th Psalm can be heeded: "They that make them are like unto them: so is every one that trusteth in them." A foundational philosophy should focus not just on science, but on all forms of human

activity. A general suspension of the *Thesis of The Natural Attitude*, to use Husserl's phrase, should be the goal so that the basis of the human experience is raised to consciousness.[6]

That a critical and reflexive effort of this type involves dangers is clear. One of the greatest is yielding to a kind of personal relativism which can lead to nihilism. If all human statements, including scientific statements, are perceived as totally relative, then the search for truth is elusive, reason is suspect, and morality is abandoned. It is because of this danger that the foundational philosophy must be truly radical so that its truths are not only critical but affirmative of man and of thought.

A radical reflexive philosophy reveals that it is man who creates his social world, who gives meaning to his experience and who interprets nature to herself. The intentional cognitive relation of man to the world is as basic as sense perception. The affirmation of this cognitive relation is an affirmation of man's "directional creator relation" described by Owen Barfield in *Saving the Appearances: A Study in Idolatry*.[7] It is the affirmation of man as the meaning giver, the creator of the world he experiences. And it is an affirmation that is objective and scientific in the sense that it is grounded in experience and verifiable by those willing to reflect without prejudice.

What this reflexive effort yields is not a totally relative relation to truth, but a tentative and historically changing relation to a world that is subtle and complex. It leads to the experience that "Truth exists despite man's inability to know the final and complete truth and from this affirmation is derived the certitude that love is better than hate, courage better than cowardice, justice better than injustice and freedom better than slavery."[8]

The image of human knowledge which emerges from such a foundation philosophy is integrative for it sees all knowledge grounded in the intentional experience of the world. History becomes the history of this intentional relation, within a given cultural linguistic framework, as well as the history of overt acts. Art, literature, philosophy, and science are different expressions of this intentional relation. Is it, for example, an accident that the development of perspective in art during the Renaissance, which implies an experience of object-subject separation, coincided with the beginnings of causal quantitative science?

Science differs from other modes of expression in that its interests are the exploration and control of the physical world revealed to the senses and that it is explicit and rigorous about its method. Its interests are therefore different from art or history, which are not concerned with the laws of the material world. Despite this difference the basis of science is the same as other forms of knowing, the giving of meaning to the world.

The scientific method has raised to full consciousness sense perception. By so doing it has hidden from view the central role which cognition plays in rendering sense perception intelligible. A foundational philosophy which makes this process explicit would begin to clarify man's central role in the world. It would generate a man-centered vision of human knowledge, without which the babble of tongues will worsen, and man will forfeit his living image.

n Thinking", 36" x 24" (Watercolor).

Three

The consequences of this view for changes in undergraduate education at M.I.T. and other universities is great. It would imply a redirection of early undergraduate education away from the acquisition of skills toward creating a living experience of the unfolding of human consciousness. Art, philosophy, mathematics, science, and political and social forms would be presented in an integrated fashion as historically, culturally conditioned expressions of the human experience. This might be accomplished in a basic freshman course, although as a general awareness it should pervade all teaching. Given the enormous amount of material to be covered in a course of this type, it would basically constitute the freshman curriculum. It could be institutionalized in a variety of ways, including lectures, seminars, and workshops. The course would have to be taught by teachers from different disciplines, who through dialogue agreed on its format and content. An effort would need to be made to recreate each cultural period in as authentic a way as possible. In learning past mathematical and philosophical forms and in observing and describing physical phenomena, certain skills would be acquired. More importantly a living experience of man and his achievements would be fostered.

A number of possibilities exist for developing the curriculum during the following years on this basis. The sophomore year might usefully be devoted to a work-study program, focusing on man in a technical civilization. The year might be organized around basic headings such as Man and Physical Environment, Man and Society, and Man and Culture. It would build on the previous year by providing an experience of our present culture. Through workshops, it would encourage critical reflection on problem interrelationships.

The junior and senior years might then be devoted to a more traditional acquisition of specialized knowledge. The course materials should, however, be presented in a developmental and integrated manner; which is to say the historical development of fields and their relation to other disciplines should be treated in more than a cursory manner. In addition, critical reflexion should be fostered, preferably in conjunction with specialized course work so that the paradigms of individual fields are made explicit and their value orientation revealed.

If properly carried out, the resulting education would be integrative and man-centered, rather than fragmented and technique oriented. It would create an awareness of how fact and value are linked. A future physicist would see how the development of physics is grounded in social life, how it, like other human activities, is reflective of the intentional relation to the world. The very nature of this realization would be a block to dogma, as well as an inspiration to creativity.

The suggested curriculum changes would require basic changes in the university. For example, the tendency toward the establishment of specialized departments would need to be reversed. Four or five divisions, such as the Life Sciences, the Natural Sciences, the Social Sciences, and the Humanities would be adequate. Although this reorganization at first appears utopian, the tendency within many academic fields is toward cross-disciplinary research, raising the possibility that old fiefdoms might be replaced by larger aggregates. In addition, an integrated view of human knowledge would, if shared, provide an impetus for such reorganization.

If the outlined perspective and the briefly indicated curriculum change has merit, what are the chances for a movement in this direction at M.I.T.? They are limited, yet there is a substantial amount of discontent among students, faculty and administrators, about the undergraduate program. Extensive experimentation is underway. The work-study option for undergraduates has been enlarged. Financial backing to undergraduates for research projects has been expanded. Various experimental educational programs have been created, providing undergraduates with considerable flexibility in meeting Institute requirements. These are the Experimental Study Group (ESG), the Unified Science Study Program (USSP), and Concourse. All three programs were designed for freshman and sophomores. They all emphasize greater student-faculty interaction, a more informal learning environment and greater integration of subject matter. Concourse, in particular, has sought to bring together in teams faculty and students with different interests, who together select themes to be studied from the viewpoint of different disciplines.[9]

Although it is too early for a thorough evaluation of these programs, the Committee on Educational Policy has recommended their continuation. Programs like Concourse and ESG are similar to the curriculum changes recommended in this essay. A guarded optimism about the eventual integration of the sciences and humanities at M.I.T. would therefore appear to be possible. However two important factors must be considered. The first is that these programs take place within an institutional setting which defines the success of these programs in terms of their ability to teach the skills otherwise taught in the basic required undergraduate courses. This places severe restrictions on the content and style of teaching. More importantly, these programs do not proceed from an explicit view of knowledge which would make the integration of the sciences and humanities natural rather than contrived. If I am correct in believing that the separation of the humanities and the sciences at M.I.T. and elsewhere reflects a fundamental crisis in the conception of knowledge, then curriculum experimentation and organizational tinkering will not be sufficient. Rather, experimentation must proceed from a different image of knowledge. I have pointed in the direction of one such image. Whether it is plausible in the long run remains to be seen.

Four

Can our culture generate a view of knowledge more capable of redirecting education and of giving meaning to a world threatened by senselessness? The pessimism of thinkers such as Jacques Ellul and Jurgen Habermas about the future of technological society seems at least partially misplaced.[10] The phenomena of the counter-culture is a testament to a search for new meanings, to a desire for alternatives to the values of efficiency, rationality and control. Because of this search and its unintended consequences, Max Weber's vision of bureaucratic society

38

moving toward "a polar night of icy darkness and hardness" is at least balanced by Yeat's antithetical vision in "The Second Coming":

Things fall apart; the center cannot hold:
Mere anarchy is loosed upon the world,
The blood dimmed tide is loosed, and everywhere
the ceremony of innocence is drowned;

A kind of tenuous balance appears to exist at the present time between these two visions as historical possibilities. The resulting freedom could be used for the creation of a new humanism, which might foreclose either historical possibility. M.I.T. could participate in such a task, thereby helping to preserve the scientific heritage in the best sense of the term. Or it could forego the challenge and continue to transform the humanities and the social sciences into specialized "scientific" disciplines according to the dictates of a technical metaphor. The first choice involves more than a commitment to experimentation; at a minimum it involves finding a suitable group of people dedicated to the search for a new humanism and providing them with an environment in which they can pursue this goal.

REFERENCES

1. M.I.T. Commission on Education, *Creative Renewal in a Time of Crisis* (Cambridge, Massachusetts, 1971), p. 80.
2. R. Dubos, *The Dreams of Reason: Science and Utopias* (Columbia University Press, 1961); J. Weizenbaum, "On the Impact of the Computer on Society," *Science* (May 12, 1972), pp. 609-614.
3. For an introduction to phenomenology see P. Thevenez and J. Edie, *What is Phenomenology?* (Chicago, Ill.: Quadrangle Books, 1962) also E. Husserl, *Ideas, A General Introduction to Phenomenology* (Collier, 1962); a useful introduction to Steiner's works is *A Theory of Knowledge Based on Goethe's World Conception* (New York, N.Y.: Anthroposophic Press, 1968).
4. W. Heitler, *Man and Science* (London: Oliver and Boyd, 1963). pp. 31-58.
5. H. Smith, "Science and the Human Spirit," *Technology Review* (Nov. 64), p. 30.
6. E. Husserl, "The Thesis of the Natural Standpoint and Its Suspension" in J. K. Kockelmans, *Phenomenology: The Philosophy of Edmund Husserl and Its Interpretation,* (New York, N.Y.: Doubleday, 1967), pp. 68-79.
7. O. Barfield, *Saving the Appearances: A Study in Idolatry* (New York, N.Y.: Harcourt, Brace and World, 1968), p. 159.
8. T. P. Govan, "The Task and Purpose of the University—A Report on the Educational Writings of Frederick Denison Maurice," *The Anglican Theological Review,* Vol. 47 (1965), no. 4.
9. Committee on Educational Policy, *Summary Statement on the Experimental Programs, M.I.T.* (Cambridge, Massachusetts, May 5, 1972); M. Parlett, *Study of Two Experimental Education Programs at MIT* (Cambridge, Massachusetts, December 15, 1971).
10. See Jacques Ellul, *The Technological Society* (New York, N.Y.: Alfred A. Knopf and Random House, 1964); Jurgen Habermas, *Toward a Rational Society* (Boston, Mass.: Beacon Press, 1971), pp. 81-122.

Improving M.I.T.
As An Environment
For Humane Learning

From *Creative Renewal in a Time of Crisis*
Report of the Commission on M.I.T. Education 1970 .

In practical terms we must face the question of why, after considerable effort to overcome their separation, we find the humanities and the sciences still not well-integrated in the intellectual life and the educational programs of the Institute. M.I.T. has always recognized that it could not responsibly offer *only* a technical training to its students but also had to provide them with a broader education. In recent decades it has strengthened the humanities and social sciences curricula and has sought in other ways to enrich its cultural life. Yet we cannot say that these efforts to make humanistic learning and concern with questions of value as important as they should be in the life of the Institute have thus far been as extensive or as successful as we would like. What is the root of the problem?

The difficulty, in our judgment, is that the general environment at M.I.T. is too narrow: it does not adequately encourage or sustain humane learning in the fullest sense of the term.

To some extent, this is a consequence of the intellectual problems we have noted. But more is involved than the kind of intellectual fragmentation and narrowing that we have described. The Institute is in fact so dominated by the ethos of science and technology that other modes of thinking and other approaches to the analysis of human and social problems do not, for the most part, receive the serious consideration they should. Despite M.I.T.'s recognition of the importance of the humanistic and social disciplines and despite its real efforts to enhance the stature of the humanities and social sciences and the arts at the Institute, these all still play a marginal role here; too many students and too many faculty members continue to think they are unimportant, irrelevant, methodologically "soft", and hence not productive of new knowledge, appropriate perhaps for those romantic periods of self- and social concern that all students seem to have to go through, but not really germane to the central concerns of the Institute.

However much M.I.T. may counsel its students about the importance of the humanities and social sciences and tell them that an education is incomplete without some experience in those areas of learning, the very structure of the curriculum and the clear requirements for success at M.I.T. encourage them to relegate such studies to a minor secondary role in their intellectual lives. They are led to recognize that certain fields, disciplines, and subjects and certain modes of analysis count and that others do not; that it is their technical proficiency above all that the Institute *really* cares about. Immense value is put on technical problem-solving ability, on the acquisition of instrumental skills; less is demanded in other areas, and the message gets through. We note, for example, that first- and second-year subjects in the humanities and social sciences carry fewer credit "units" than science subjects do — an almost trivial example but one that points, we think, to deep-seated assumptions and values at M.I.T. that manifest themselves in a variety of ways tending to narrow our intellectual environment.

The dominant technical, quantitative style of the Institute operates to stifle concern with other modes of knowing and expression and tends to divert attention from aesthetic and other non-quantifiable values, as is evident by a glance at our extremely functional and unlovely physical environment, which in many ways is brutally austere and ungraceful. Inadequate attention is paid to the affective aspects of learning, and the arts still play a very small role in our total experience at the Institute. Visual education, despite some remarkable efforts in this direction, has been neglected; that there is a craving for more beauty and color in our community is suggested quickly enough by a walk through the main hall of the Institute, whatever one may think of the wall-art that has thus far appeared. The Kresge Auditorium, the M.I.T. Chapel, and the Student Center have done a great deal to humanize the student environment, but much more remains to be done along the same lines.

In the most general sense, despite all the changes of the past decades, there remains at M.I.T. a decided bias against humanistic learning. How this can be modified is a difficult question, but it is one that the Institute cannot avoid. The problem involves more than the relegation of the humanities and social sciences to a minor (and in some measure avoidable) part of the curriculum; it is a result also of the overwhelmingly skill-oriented, problem-solving approach of the scientific and engineering education offered at the Institute. No one would for a moment deny the great importance of acquiring and mastering the analytic skills required for competent work in the scientific and technical disciplines and fields. We would insist as strongly as anyone that students still be asked to achieve in reasonable measure this kind of competence; what we do question is the wisdom of allowing this unimpeachable aim to prescribe nearly the sole mode of learning what science and engineering is all about. Studied in a certain way, scientific and engineering subjects can be culturally broadening — such subjects too, after all, have a humanistic dimension and speak to the largest intellectual and spiritual aspirations of men, as has been pointed out by many of our forerunners at the Institute, from William Rogers to Henry Pritchett to Karl Compton.

But these dimensions of science do not receive the emphasis they might have in the Institute's educational program; and we do not accept the argument that this cannot be done without corrupting the technical quality we want to maintain in our science and engineering education. Our students and faculty should be asked to understand and to be aware of the social, the philosophical, the "human" dimensions of science and engineering. The present virtues of our programs, which are admirably concrete and practical, which emphasize the *experience* of *doing* science and engineering, and which thwart in some ways an over-abstract or over-intellectualized approach to such studies, should be maintained; but such virtues, if not balanced by reflection about the history, the assumptions, and the intellectual and social purposes of those fields, as well as by some understanding of the social contexts in which scientists and engineers work, tend to become limitations.

We should try to develop instrumental skills *and* understanding of the philosophical and social dimensions of science and engineering; to some degree this can be accomplished within the scientific and technical subjects and courses themselves. The task cannot rest with the humanists and social scientists alone — if only because their training has too often made them illiterate about science and technology. The prevailing narrowness of approach to the study of science and engineering can, we think, be modified without running the danger of producing incompetents. It is entirely possible that by broadening its style of science and engineering education (again, without abandoning the goal of technical proficiency) M.I.T. will stimulate the creative potential and the scientific and social imaginations of its students to a greater extent than it has thus far. It is not only by genuinely enlarging the role of the arts, the humanities, and the social sciences at M.I.T. that we can create an environment more hospitable to humane learning; we can do so also by encouraging a broader view of learning and a deeper engagement with questions of value in the scientific and technical disciplines themselves.

Other aspects of the environment also call for comment. The prevailing values of this institution place very great stress on productivity, efficiency, action, organization — to the detriment of more contemplative, casual, and spontaneous modes of intellectual life. Certainly these former qualities are great virtues, but in excess they work against reflectiveness about purposes and values and tend to make difficult the kind of ongoing self-consciousness about the larger meanings and consequences of one's work that we always need. M.I.T. has the appearance of great business and terrific efficiency, but not of great reflectiveness — an impression conveyed by more than one visitor to this university. The very fact that we have "awakened" to find ourselves uncertain of our purposes and direction lends weight to this judgment. It would be easy to caricature what we have in mind; the point is difficult to make but we feel it must be made. There is a relentless, driven quality to life at M.I.T. that leaves little room for quieter intellectual activities, for the kind of moderate slackness that is often a condition of creativity and genuine communication. To some extent this is inescapable; we must accept the paradox that the serious pursuit of knowledge requires in some measure a narrowing of attention and a concentration of energy; that a certain intensity and single-mindedness are necessary conditions for intellectual discovery and intellectual excellence and surely for productiveness. But these are not conditions we have to worry about creating at M.I.T. Rather the opposite is true. We need to think about ways of somewhat slowing the pace of life here, the better to evaluate our activities and reorient, when necessary, our search for new knowledge by taking into consideration more consciously than we now do the ends we wish to serve or fulfill.

We have in fact already taken sensible steps in this direction — the freshman pass-fail and calendar experiment among them. We should also consider other ways of reducing the work load at the Institute and of increasing opportunities for individuals to pace their work more in keeping with the rhythms of their inner lives. We have no intention of proposing that M.I.T. adopt as an ideal an image of lolling ease and self-indulgent disorder; order and method are among the great and valuable achievements of civilization. But one must breathe. One wonders how much a student taking five or six subjects really is learning.

In many ways M.I.T. is a fragmented community. This is yet another aspect of the environment we must consider. The dominant ethic at the Institute is one that urges individual striving and individual excellence — this despite the tradition of collaboration and sharing that is so strong in science and engineering. To be sure, much collaboration does go on at M.I.T. — for example, many projects are genuinely collective efforts. In both oblique and sometimes systematic ways we do learn from one another and assist one another . Yet the Commission feels that there may be too much stress on individualism and not enough on a community of effort and purpose at the Institute. This is, unfortunately, especially true of students at M.I.T., who, even though there is a visible reaction against competitiveness among them, still strive so hard to excel as individuals that they do not adequately learn what genuine cooperation is. The result is the perpetuation of a kind of isolation and alienation among them that makes the experience of other perspectives and values difficult. The strong individualism that is a result of achievement-orientation works to diminish a sense of community and reduces the opportunities for a kind of casual, humane learning — about how others think and feel — that no curriculum can produce. The more we are isolated from one another, the less adequately we know what it means to be fully human. This is true not only of the students at M.I.T. The faculty members are submerged in their work, separated by their functions, isolated by rituals and status arrangements, kept apart by professional preoccupations, imprisoned in specialized languages, scattered throughout an enormous academic city. They are turned in on themselves in a variety of private enterprises. Even physical conditions and arrangements are not conducive to a strong sense of community: many students and most of the faculty of necessity live at some distance from the Institute; and the various departments and disciplines at the Institute are not only intellectually but physically separated. These are problems for which there are no clear answers and for which in some instances there are no answers at all. But the real fact of our fragmentation is something we ought constantly to bear in mind and try to overcome in whatever ways seem feasible.

One possibility that some members of the Commission feel is worth exploring would be to experiment with small learning "communities" of fifty to one hundred students, each with its own faculty and even with its own physical center.

Though the effort would be immense and the departure from current operating however difficult to arrange, the gain in our view might be significant: communities of shares purpose provide opportunities for liberal and humane learning that are important for the broad education of the students and for the strengthened engagement of the faculty in the life of the Institute.

In our judgment, it would also be a good idea to organize a month-long, 'Institute-wide conference on the subject of "Knowledge and human Vaues" to take place as soon as it can reasonably be arranged. During the conference period, scholars and social critics with different views and perspectives should be invited to participate in formal and informal meetings with M.I.T. faculty and students. Topics might include the values or consciousness of the "youth culture", the nature of scientific reasoning, and the conflict between technocratic and democratic decision making. The conference should also consider ways of making discussions of this kind a regular feature of M.I.T. education. If well designed, such discussions could add a significant new dimension to the intellectual life of the Institute and offer a badly needed opportunity for communication on problems of general concern.

We realize that the phenomena we have been discussing are simply heightened local manifestations of general cultural tendencies and values in our society, which is fact-minded, pragmatic, individualistic; preoccupied with technical progress and power; restlessly active; and still dominated by the work ethic. Nationally, we have only recently achieved awareness of the need to reexamine our purposes and to pay much greater attention to the quality of our lives. In a sense, we are suggesting that the Institute make an effort to transcend the surrounding culture; if this cannot be accomplished in our universities, then where will it be done?

Science and Technology in the Service of Man

In a certain very real sense, what we need to do to improve the environment at M.I.T. is similar to what needs doing in the larger society. Great achievements have been recorded in science and technology, but as we consider the current state of the world, we cannot fail to be struck by the disparity between the level of our accomplishments in science and technology and that of our attainments in improving the quality of human life.

Who is to blame for this disparity? Some claim it is human nature which leads us to seek power through knowledge only to become the victims of our own reckless pride and ambition. Others say it is our acceptance of a "technological imperative" that holds in effect "whatever can be done must be done." Others blame our social systems and ideologies. If there is no shortage of culprits, however, there *is* a shortage of constructive suggestions as to how the situation can be improved.

We at M.I.T. have always believed that science and technology can have enormous benefits for civilization. It is true that many of the social problems we face today are by-products of advances in science and technology, but we must bear in mind that in a great many cases these problems have arisen only because the human race has managed to solve earlier challenges to its survival and evolution. If we can worry now about overpopulation, it is because we have conquered many diseases and are able to prevent many of the premature deaths that previously checked population growth. If we need better systems of transportation, it is because the barriers of distance have been dramatically reduced and new expectations have arisen with respect to mobility and ease of communication. If we can be concerned now with providing everyone with decent housing, equal educational opportunity, and a chance to build a career rather than simply perform a job, it is because we have so vastly improved the productivity of labor by advances in technology that there is no longer any need to accept the proposition that most people must be condemned to a life of drudgery and social inferiority.

It is too easy to lose sight of the great progress that has been made, thanks to the advance of science and technology, and to blame scientific reason for failings due far more to misdirected human and social passions. But it would be a grave mistake for anyone to think that the problems we now face can best be treated by curbing progress in science or by somehow turning off the technological tap. Professor Victor Weisskopf, who shared with the Commission his concern for the future of science, quoted the words of warning of the philosopher of science Michael Polanyi:

> Encircled today by the crude utilitarianism of the Phillistine and the ideological utilitarianisms of the modern revolutionary movement, the love of pure science may falter and die. And if this sentiment were lost, the cultivation of science would lose the only driving force which can guide it towards the achievement of true scientific value. (*Personal Knowledge,* University of Chicago Press, 1958)

With Professor Weisskopf, all of us at M.I.T. believe deeply in the value of science for its own sake — as an expression of the age-old human desire to know, and through knowing, to be free of fear, ignorance, bias, and superstition.

We also value science because we know that it is indispensable to progress in technology. We value technology because we believe that useful knowledge is indispensable to social progress. If we are to save the natural environment at the same time that we extend our productive capacity to assist those who still endure poverty, we will need to develop sophisticated systems of industrial management. We will need to create early warning systems to detect dangers before they become too difficult to manage. If we are to cope with decay in the cities, we will need better and cheaper forms of public transportation, newer techniques in construction and housing, better systems for the delivery of health care. Achieving these improvements will require the best talents and best resources universities can offer.

Perhaps the hardest set of problems society faces are those of the environment. We have several times called attention to the unfortunate consequences of the separation and distillation of knowledge into separate disciplines. Nowhere are these consequences more evident than in the crisis of the environment. The failure to examine our collective behavior in the context of a single, finite, interactive system has led us to the brink of catastrophe. Solving the problems of the environment will require more than simply substituting non-polluting for polluting automobile engines. We must also consider the habits of mind and the cultural values that have led us to worship consumer goods and desecrate the environment which gives us life and connects us to the rest of the natural universe.

Technology alone will not solve our social problems, but it is a critical instrument of any constructive solution to many of them. In the past, societies generally devoted technology to the improvement of production and to raising the standard of material comfort. There is still much to be accomplished in these areas. Now, however, the critical resources of technology, human and material, must be turned toward a set of new social concerns. In the "post-industrial" stage of history, our continuing preoccupation with overcoming material scarcity and curbing the dangers of the natural environment must be balanced by a heightened concern for the quality and purpose of individual and social life. That is one reason why it is so important that M.I.T. education be redesigned to make possible a broader and more responsible kind of professionalism. It is also why we must begin to devote more of our resources to the identification of major social priorities and to attempts to assist governments and groups of concerned citizens to act on these priorities.

Public service has always been a major concern of M.I.T. Today, the concern for public service must take the form of a new effort to cope with pressing problems that have been badly neglected until now. In the decades ahead, the pressures of unchecked population growth, the depletion of mineral resources, the alarming deterioration of the environment, and many other problems will call for the highest possible degree of technical and scientific sophistication. Technologists will be called upon to find temporary as well as long-range solutions to such problems, in order to allow society the time it may need to develop more fundamental approaches.

We do not propose, therefore, that M.I.T. in any way abandon its fundamental commitment to the application of science and technology to the work of society. M.I.T. must continue to be a place where the rational formulation and solution of problems is the leading concern, where the phrase "scientific method" describes a serious effort of scholars to understand man and nature, and not the misuse and misapplication of scientific methods, which may properly be criticized. No complex society today can survive, let alone solve its pressing problems, without the help that engineering and the engineer's understanding of what "process" is about can provide. The changes in educational format which we introduce at M.I.T. must be designed to preserve that critical aspect of M.I.T.'s fundamental structure that is aimed at building from the sciences — natural and social — to the technologies whose ultimate purpose is the service of man and society. We must make a similar effort with respect to public service.

We therefore propose, first, that M.I.T. identify problems of major social concern in which it can usefully apply its resources — problems such as pollution, the urban environment, transportation systems, arms control, and the development of new medical technologies and delivery systems. Second, M.I.T. should develop capabilities designed to assist public agencies in assessing the costs and benefits of new technology and in evaluating the effects of the transfer of technology from advanced to less advanced societies. The subjects should include examinations of the effect of technology upon public health and population growth and the social consequences of various systems of transportation. Third, M.I.T. should direct research and resources not only to national and international problems but also to the needs of its immediate regional area. Finally, M.I.T. needs to continue its development of environmental quality research programs, as suggested by the Environmental Quality Task Force. This effort should be framed in such a way as to promote participation by students and faculty in active efforts not only to study the problems but to assist in their solution. The programs should also be designed to include the participation of faculty and students from many disciplines, including management and the social sciences.

In order to facilitate the accomplishment of these goals, we believe it would be advisable to develop an administrative focus for the Institute's public service efforts. This office could be charged with the responsibility for coordinating the application of Institute resources to major problem areas and for surveying changing tendencies in the area of public needs.

In making the above recommendations, we are deeply conscious of the fact that the principal way in which the Institute serves society is, and should remain, through innovation and leadership in education and research. All of our proposed public service efforts should be reviewed as to their impact on our basic educational enterprise and the manner in which they contribute to the environment for learning at M.I.T.

Future Directions for Technology Assessment

Vary T. Coates
Head, Technology Assessment Group
Program of Policy Studies in Science and Technology
George Washington University

Technology assessment is the systematic identification, analysis, and evaluation of the potential impacts of technology on social, economic, environmental, and political systems, institutions, and processes. It is concerned particularly with the second and third order impacts of technological developments; and with the unplanned or unintended consequences, whether beneficial, detrimental, or indeterminate, which may result from the introduction of new technology or from significant changes in the application or level of utilization of existing technologies.

Technology assessment seeks to identify options and clarify tradeoffs which must be made, by anticipating in so far as possible societal and industrial changes which may be set in motion by a decision or an action. This approach is designed to provide an objective and neutral informational input into decision-making and policy formulation with regard to science and technology. The analytical techniques of technology assessment may be integrated into the on-going process of planning, designing, and evaluating technological projects and programs, and may also provide an external review and evaluation of such projects and programs at any point in time.

If one assumes for the moment that it is possible to anticipate and at least in a crude way estimate the magnitude and importance of secondary and higher order impacts, why should this be done? Clearly technology can and often has set in motion processes of social change which eventually affect whole societies. The mechanization of agriculture, for example, followed by the so-called genetic and chemical revolutions in agriculture, released millions of rural workers from the land only to send them pouring into the cities, which were unprepared to receive and absorb them. John Eberhard has noted that the cities themselves, as we know them, are the products of six inventions which occurred in one eleven-year period (1877-1888): the electric trolley, steel-beam construction, elevators, electric lights, the automobile, and the telephone. Some, including myself, would argue indeed

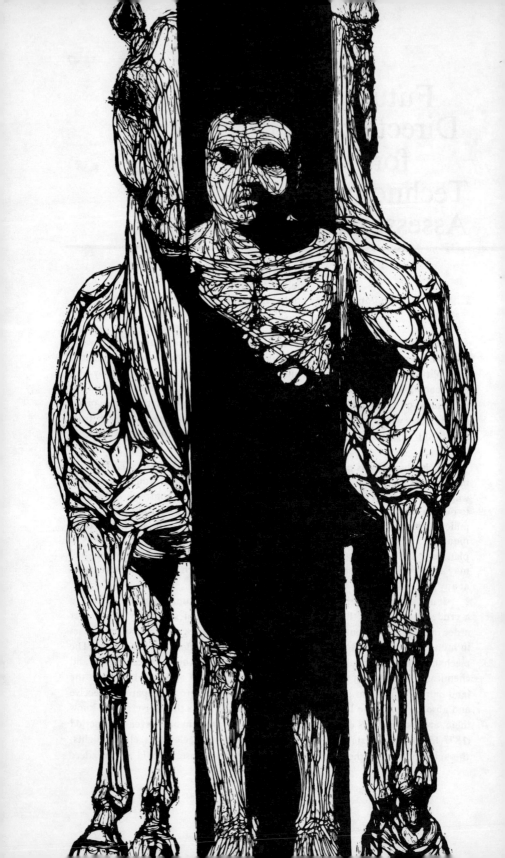

that technology is *the* basic causative agent in social change. We speak of social change only in regard to man, because in the animal kingdom all change is evolutionary; that is, in response to the changing environment. It is only Man the toolmaker who changes and modifies his environment and thereby sets into motion changes which he may not be able to foresee or control.

Not all technology is, of course, hardware since technology is "the systematic application of scientific knowledge to the solution of a problem or to modification of the environment." Social technologies also have secondary consequences and unplanned impacts; consider, for example, the invention of the credit card, or the Homestead Act, or the G.I. mortgages which gave rise to booming suburbs after World War II. However, in the remarks which follow, I am for the sake of simplicity thinking mostly of technology as the use of industrial and engineering tools and processes.

What are the consequences of performing technology assessment? Programs, projects, and products are usually evaluated through some form of cost/benefit analysis, but both costs and benefits are generally very narrowly construed. It is ordinarily the direct costs and the direct, immediate benefits to the sponsor or investor, and to the user, which are considered. But if we agree that technology drives, or helps to drive, social change, this creates the opportunity for alternative lines of technological development and hence alternative societal futures. In other words, we may be able to invest public and private resources selectively in those lines of technological development which will maximize societal benefits, and at the same time have the possibility of avoiding, minimizing, or controlling the undesirable aspects which, in the human condition, accompany even the most desirable action or decision. There are always tradeoffs to be made. Thus, anticipatory technology assessment offers a number of possibilities:

Desirable technological developments and applications can be encouraged, stimulated, fostered, or enhanced, through a variety of incentives including in some cases investment of public funds in exploratory R & D;

By detecting unsuspected or unplanned potential benefits, new applications of existing technology can be encouraged;

When potential impacts are uncertain or impossible to fully evaluate (for example, if the eventual level of utilization is problematical), monitoring may be called for to provide early warning of detrimental effects;

Where detrimental impacts are inevitable, but outweighed by potential benefits, controls can be instituted, such as pollution limits, licensing arrangements, or safety regulations;

In some cases projects or programs can be modified, relocated, or redesigned; and finally,

orseman", 5' 1" x 2' 6" (Woodcut).

Unjustified, undesirable, or excessively dangerous projects can be blocked before heavy resource investment has been made.

How does one do technology assessment? There is, unfortunately, no magic formula and no universally accepted systematic methodology. The essence of technology assessment is to define the concepts associated with a new technology or an innovative application of technology and to imagine its potential impacts, attempting to evaluate their probability, significance, direction, magnitude, and duration; to identify wherever possible those which indicate the need for either avoidance or enhancement; and to generate policy options for the use of the decision-maker. The classical statements on the methodology of technology assessment were two 1969 reports of the National Academy of Sciences and the National Academy of Engineering.[1] Basically similar, they can be reduced to a structured analytical process involving several simple steps or tasks:

Description of the subject technology and its parameters and development of the data base;

Description of alternative, supporting, and competitive technologies;

Development of state-of-society assumptions, for present and future time periods;

Preliminary identification of potential impacts by considering each aspect of the development: extraction and utilization of raw materials, investment of capital and labor, industrial processes and their emissions, land use, transportation of products, utilization by consumers, and effects on other goods and services;

Analysis of impacts in terms of affected parties and processes, probability of occurrence, and direction, magnitude and duration of induced changes;

Identification of possible policy options;

Assessment and comparison of alternative action options.

Technology assessment is most adequately performed by interdisciplinary teams using a variety of analytical techniques to accomplish these tasks. Obviously, assessment will draw heavily on the methods of systems analysis and operations research, although going far beyond the aspects of efficiency, performance, cost, maintenance, and reliability which are stressed by these techniques. Other techniques which can be utilized are modeling and simulation and forecasting methods borrowed from futures research, including Delphi, morphological boxes, and scenario construction.

The two general approaches are (1) a case study of a particular project, product, or application, and (2) modeling and simulation. Case studies have the advantages of greater comprehensiveness and relevancy, and generally greater understandability and acceptance, but produce results which are difficult or impossible to transfer to other projects. Modeling produces more transferable results, but is more difficult, abstract, and tends to include only impacts which are readily quantifiable.

Technology Assessment in Government

Technology assessment as an input to decision-making is applicable and appropriate both in the public sector and the private sector. Up to now most of the activity and discussions of technology assessment in the United States has focused on governmental decision-making, and that aspect will be covered first, because it has an important bearing on the questions of whether industry should, can, or will integrate technology assessment into decision-making in the private sphere.

In 1966, Representative Emilio Daddario, then chairman of the Subcommittee on Science, Research, and Development of the House Committee on Science and Astronautics, proposed the establishment of an Office of Technology Assessment to serve the Congress. This occurred at a time of rising public alarm over alleged hazards to life and health resulting from contamination of the environment by by-products of chemical and industrial processes. Moreover, large public projects such as highway and airport development had caused numerous public protests, demonstrations, and legal actions resulting in costly delays to many such projects. Growing hostility to technological programs aroused political pressures which intensified Congressional suspicion of the process of planning and programming in executive agencies to provide Congress with adequate information about the impacts of governmental programs.

During the ensuing six years the concept of the technology assessment was picked up and elaborated in dozens of professional meetings, academic conferences, scientific societies, and international meetings. Studies were commissioned by the House Committee from the National Academies of Sciences, Engineering, and Public Administration.[2] A great number of books, articles, and professional papers have been written, and a number of "methodologies" and "approaches" have been offered.

In October of 1972, a bill was passed and signed establishing an Office of Technology Assessment to serve the Congress. Six Senators and six Representatives, three from each party, have been designated members of the Technology Assessment Board, which will direct the office. This was a decided and perhaps unfortunate change from the Board as originally planned; in the original Daddario bill, the Board included officials, such as the head of the Library of Congress and the Comptroller-General, and some public members. The purpose of the change was to increase the independence of the Office from the Executive Branch and insure that it was wholly responsible to Congress. The danger in the present arrangement, which makes the Board essentially a joint committee, is that the Office of Technology Assessment may find itself constrained by politics. The freedom to foresee and choose those prospective developments, critical programs, and projects which should be assessed and the ability to carry out comprehensive assessments, may be prejudiced by the fear of telling Congress what it doesn't want to know, or of casting doubts on a program or project that a powerful Congressman has worked for years to develop and promote. Much will depend on the sensitivity and commitment of Senators and Representatives who make up the Board and on the courage and integrity of the man who will head the Office.

The Office of Technology Assessment will not itself carry out the assessments but will contract for them with various research organizations. Assessments can be initiated by the Board or by the request of the Chairman of a Congressional Committee. A Senator or Congressman has many decisions to make every day. This raises several questions. Will the Office insist that assessors outline for the decision-makers a number of alternative policy formulations? Will it provide an interpretive link between the scientific report which embodies the assessment results and the ultimate decision-maker? Or is the Congressman simply to be supplied with a bulky report and left to ferret out its policy implications for himself? How much weight, and how much credibility, will Congress accord its work? Again, the answers to these questions remain problematical, and much will depend on the respect and prestige which the Office can establish in its first year or two. Senator Edward Kennedy will be the first Chairman of the Board; the Director will probably be chosen sometime in the early months of the new Congress.

As usual, when Congress acts, the Executive Agencies react. Some are already moving to set up assessment groups or offices. But many of the pressures which moved Congress have acted on the agencies as well, and most agencies have, within the last five years, improved and broadened the process through which they plan and evaluate technological projects and programs. The lack of generally accepted methods for integrating such considerations into administrative decision-making and into the justification of agency programs, the lack of sustained impetus and encouragement from the upper levels of the Executive Branch, and the institutional deficiences which every bureaucracy builds up, cause this improvement to be slow and not uniform across agencies.

In a recent survey[3] I looked at eighty-six offices in the civilian agencies whose responsibilities were primarily technological. I concluded that 13% frequently perform or sponsor technology assessment and consider that as their major task. At the other extreme, 24% were concerned only with primary performance characteristics and with direct dollar costs. The remaining 63% do some technology assessment; most are partial or narrow assessments which take into account some secondary impacts, usually environmental or secondary economic impacts. Assessments are often viewed as support for agency planning and programming or as support to substantiate basic or applied research programs. I will discuss some of the findings of this survey, because many of these observations would also be applicable to technology assessment within a corporation, bureaucracies being much alike wherever found.

Thirty-five percent of the offices doing technology assessment said that most or all of the work was done in-house. The remainder preferred contractor studies or used a mix of in-house and contract analysis. In-house studies have greater credibility for agency management and show greater likelihood of producing institutional change; the data base remains available to the agency; additional expertise is developed, and the assessment activity can be fitted in with the time, resources, and workload of the agency. However, in-house studies have some disadvantages: most offices lack the necessary multi-disciplinary staff; there is the possibility of institutional bias, and because administrators can suppress or distort findings which displease them, internal assessments may lack external credibility.

If assessments are farmed out to independent research groups there may be less institutional bias and greater objectivity, and thus greater external credibility. More disciplines can be mobilized than a government office is like to have on tap, and the regular work of the agency staff can proceed without interference. On the other hand, contractors, too, may tend to tell the agency what it wants to hear (as they perceive it) and contractor reports can also be ignored or suppressed by agency management. Moreover, there are usually severe difficulties of coordination and management of the assessment process where there is geographical distance between the agency and the assessors. Those agencies who do use contractors or grantees for performing technology assessment tend to prefer independent research organizations over university groups, which may have difficulty in organizing a management structure for large inter-disciplinary research projects.

Government agencies which perform and sponsor technology assessments generally use engineers, economists, and physical scientists, particularly the first two categories, since these are the people they have on hand. There are few social scientists working for the technologically-oriented government offices. Independent research organizations and universities groups are more likely to include social scientists, but even here the number is usually small compared to the overall size of the inter-disciplinary team.

Most technology assessments rely heavily on the collation and judgmental analysis of existing information. Field studies, modeling and simulation, and other techniques used by economists and engineers are also used.

Analyzing 97 examples of technology assessment, that is studies which were chosen by agency officials and given to me as examples of their technology assessment work, I found that only nine, or less than 10%, could reasonably be classified as *comprehensive*—the criteria being that they were open-ended in the sense of trying to identify all significant impacts, that they were multi-disciplinary, that they were intended to support and inform public decision-making or policy formulation, and that they were funded at a level sufficient for in-depth analysis. Examples of these are the Jamaica Bay—Kennedy Airport Study, The Northeast Corridor Transportation Project, A Study of Medical, Ethical, Economic, and Psychological Implications of Cardiac Replacement, and a Study of Snowpack

Augmentation (Weather Modification) in the Upper Colorado Valley. Such studies were usually initiated by Congress or by a source other than the operating agency (e.g., the National Science Foundation). They had an average of four or five disciplines represented on the analysis team, and took about 16 months from initiation to final report. The average cost was $381,000, but the median was much less, about $149,000. A reasonable minimum for a comprehensive assessment of this sort is probably $100,000.

The most significant aspect of these comprehensive studies was that the subject matter was usually broadened or restructured during the course of the analysis as unexpected findings and considerations emerged. The recommendations which came out of the analysis were generally one of four kinds: new basic or applied research priorities, specific policy formulations, modification of accepted practices or projects, or termination of the projects. In most cases some administrative or legislative action resulted from the assessment, ranging from informal but real changes in practices to outright termination of two projects.

The remaining 86 studies were of several kinds. The bulk of them, 40 studies, were "partial" or "narrow" assessments which considered a small number of preselected impacts, usually environmental or secondary economic effects, and were not intended to be comprehensive. Fourteen were "problem-oriented" assessments. Instead of beginning with a prospective technology, they focused on a societal problem such as pollution or regional underdevelopment, in which technology is a contributing factor or to which technology may offer a solution. Fourteen studies were Environmental Impact Statements, prepared as required by the National Environmental Policy Act of 1969 for every federal action which "significantly affects" the environment. Seventeen studies were "futures research" dealing with trends affecting future utilization and development of technology, such as supply and demand studies, technological forecasts, or long-range planning documents. The three remaining were two methodological studies and a survey of assessment activities.

Narrow assessments were usually self-initiated by the agency and were regarded as part of the ongoing substantive research programs or as general informational support for agency programming. There was an average of two disciplines per team, usually economics and physical or social sciences. The average cost of partial assessments was $139,000 if done by independent research companies and $85,000 by university groups, which had a cost per professional manyear about half of that of the independent groups. (No cost estimates are available for in-house studies, since agencies do not keep accounts in that way.) University assessments took an average of 13 months, independent research organizations an average of 22 months.

Problem-oriented assessments were usually initiated from outside the agency, often by Congress; agencies rarely initiate exploratory investigation of societal problems. Less than a third appear to have resulted in traceable administrative or legislative action. Such studies usually represent a preliminary evaluation of the magnitude of a problem and their influence may be slow to mature. Problem-oriented assessments were costly, about twice as expensive as comprehensive technology assessments, perhaps because most used large research teams with an average of six disciplines per team. They took about as long as partial assessments, from a year to 18 months.

Environmental impact statements may also be considered as partial technology assessments. They probably cost less than other partial assessments; a fair estimate is $10,000 or about 3 man months for the average impact statement, of which about 200 are prepared per month. The National Environmental Policy Act, by forcing agencies to collect information, by providing experience in multidisciplinary consideration of secondary impacts, and providing a mechanism for public review of decision-making, has created and maintains a strong stimulus to further development of technology assessment capability.

Since technology assessment deals with anticipation of impacts, it is necessarily future oriented. The seventeen Futures Studies were, however, concerned with trends influencing the future levels of utilization of technologies—supply/demand projections, technological forecasts, and long-range planning studies. Only one attempted systematically to lay out alternative socio-political scenarios for the future. A variety of analytical techniques were used, including trend projection and extrapolation, survey research, and Delphi. There was an average of two disciplines per team, usually engineering and economics, and the costs varied widely. Government officials consistently complained that the civilian agencies had lagged behind industry in developing technology forecasting capability.

On the basis of this study, I recommend that Congress and the Office of Management and Budget should provide additional resources and strong directives for expanded technology assessment and futures research. Emphasis on performance of technology assessment should not wait on development and acceptance of new methodology, which will come from experience and experimentation in conducting assessments. The demands made by these agencies should be substantive rather than procedural. I do not believe it is desirable to have a formal requirement for technology assessment statements on the model of environmental impact statements. A source of independent technology assessments should be provided. Maximum objectivity and usefulness to public decision-makers can be achieved if assessments are sponsored by a federal entity having no responsibility for the project or program to be assessed, and are conducted by independent research groups whose objectivity and neutrality is their most valuable stock in trade. Finally, I recommend that the National Science Foundation sponsor a survey or search to identify all potential technological innovations for which anticipatory assessments are needed and to develop priorities for allocating the resources which are available for technology assessment. Such an effort is now getting underway.

Technology Assessment and the Corporation

Should business also do technology assessment? A number of large corporations are investigating the use of technology assessment as part of their corporate planning; many have bought my report on the Executive Branch and have asked for advice in setting up technology assessment functions. My impression is, however, that there has been more talk than action so far, and that is not surprising. The first business of business is to stay in business, i.e., to make a profit. Business, unlike government, does not have the broad charter to promote the general welfare, and technology assessment is in the narrow sense "non-productive" work. On the other hand, some businessmen have been heard to argue that profit-making is a necessary constraint but not the primary objective of business, which is to deliver useful goods and services to society and to make the benefits of technology fully available to a diverse, pluralist society. For the moment, at least, the corporate image requires a display of social responsibility, environmental concern, and a role in the community. Corporate managers today (perhaps lacking the cut-throat drive of old style entrepreneurs) are often men of social conscience and are sincere in this intention.

A more urgent reason for considering technology assessment is obviously that the public, and therefore the government, may really insist in the future on asking hard questions and may even be prepared to make hard choices, if the environmental degradation, the energy "crisis", and the limitations on space and natural resources become more acute. It is obviously in the interest of business to anticipate and provide the know-how and technology for utilizing scientific breakthroughs which offer new societal benefits or new approaches to solving old problems.

Business will have most of the problems and constraints of government, and some additional ones, in seeking to perform technology assessment: lack of time, funds, and interdisciplinary expertise; bureaucratic inertia and fear of rocking the boat; the problems of justifying additional "soft" research when the potential payoff may be years in the future; the lack of universally acceptable and generally applicable methodology. In addition, corporations have the problem of making assessment results acceptable and understandable to the top echelons of management and more particularly to their stockholders.

I would advance several propositions for consideration, as to how corporations should proceed in developing technology assessment practices. First, assessment must rest firmly on a well-developed capability for technological and social forecasting and a strong future orientation. Secondly, assessment teams must learn to tap the information and insights afforded by social scientists as well as those of engineers, economists, and management experts. The reward system must encourage each corporate planner and assessor to be multi-disciplinary in his orientation, rather than narrowly specialized, to be sensitive to the innovative and speculative as well as the safe conventional wisdom of his own discipline, and most of all to be creative and daring. A high failure rate, in basic and applied research and in forecasting as well, is tolerable and even necessary, because it is the cost of encouraging experimentation and innovation.

Thirdly, the assessment function, however it is organized, should be located near and directly responsible to the ultimate decision-maker in the corporation, because acceptance at the top level is necessary for effectiveness. But acceptance does not mean telling the decision-maker what he wants to hear. He must rather welcome the opportunity of being told at least part of the time what he particularly does not want to hear. It is, I think, a fundamental error to give the assessment function to the operating agent who has a professional and a personal investment in protecting and nourishing his programs, products, and practices.

The most important ingredients in business technology assessment as in government, will be foresight, sensitivity to social change, initiative, and daring.

It is engineers who design the technology and must provide the technical information on which anticipatory assessments must be based. Beyond this, it is the physical scientists and the engineers who can be expected to foresee the directions in which research and development is leading and the new possibilities that will open up. They must, therefore, provide an early warning system long before society in general can be aware of potential new benefits to be reaped or impending dangers to be avoided. Engineers can provide this service only if they are trained to be alert to the social implications of technology and aware of the complex problem of social change. At a minimum, I suggest, engineers should study the history of science and technology.

Technology assessment teams are, or should be, first of all interdisciplinary. They should include ecologists, physical scientists, economists, demographers, urban planners, and most importantly, social scientists.

In the 86 Federal offices which I studied and the nearly one-hundred assessments I looked at, the disciplines most commonly used were engineering and economics. There are several points to be made here. First, engineers must be trained to do interdisciplinary research—not simply to do engineering as part of a project that also employs other disciplines, but to exchange meaningful information across disciplinary lines, to absorb insights from other disciplines, and to integrate and synthesize information of many kinds. Secondly, the engineer must learn to deal with uncertainty. There are few general theories or laws in social science. Engineers appear to be uncomfortable in dealing with the low-probability predictions which social scientists must use, and therefore they complain that social scientists cannot tell them what they want to know. Yet social scientists might, for example, have warned highway engineers about the anomie and alienation which might result from dislocating a settled functioning ethnic neighborhood.

The manager of an assessment team is quite likely to be an engineer. He then has the responsibility of recruiting and organizing the interdisciplinary team. Social scientists, and the pseudo-sciences such as urban planning, aesthetics, etc., are still at the stage of constructing their disciplines; the reward structure is such that they are more likely to be involved in building theoretical constructs and accumulating data than to be involved in practical problems. Hence, the engineers must perhaps play a tutorial role, instructing other disciplines in the practical needs of society and the real-world constraints on meeting and managing those needs.

But the fact remains that most technology assessments today are not performed by interdisciplinary teams. They are perforce carried out by one person or a small group, and often enough this responsibility is laid on the design engineer. The engineer himself, then, should be interdisciplinary. Engineers must become sensitized to the intricate relationships and uncertainties of environmental, ecological, political, and social systems. They must inquire into the process of social change, and must be encouraged to ask not only how, but why, and what then?

Finally, future engineers must be involved to a much greater extent not only with the problems of industry and the government, but with the problems and activities of the community and the society. If the mass of citizens is to be able to understand and participate in the issues and decisions which shape their lives, rather than lapsing into obstructionism or apathy; if we are to avoid the excesses of neo-Ludditism on the one hand and technocratic elitism on the other, here too engineers must play a tutorial role. They must find ways of sharing their expertise and their understanding of the physical world with the public, who, after all, must remain the final check on decision-makers.

REFERENCES

1. *A Study of Technology Assessment.* Report of the Committee on Public Engineering Policy, National Academy of Engineering, to Committee on Science and Astronautics, U.S. House of Representatives (Washington, D.C.: Government Printing Office, July 1969) and *Technology: Processes of Assessment and Choice.* Report of the National Academy of Sciences to the Committee on Science and Astronautics, U.S. House of Representatives (Washington, D.C.: Government Printing Office, July 1969).

2. *A Technology Assessment System for the Executive Branch.* Report of the National Academy of Public Administration to the Committee on Science and Astronautics, U.S. House of Representatives (Washington, D.C.: Government Printing Office, July 1970).

3. Vary T. Coates, *Technology and Public Policy: The Process of Technology Assessment in the Federal Government.* Washington, D.C.: The Program of Policy Studies in Science and Technology, The George Washington University, July 1972. This report can be ordered from the National Technical Information Service, Washington, D.C. Nos. PB 211453, PB 211454, PB 211455 (Volumes I and II, Summary) for $19.50.

Technology Assessment: Objectivity or Utopia?

Michael Gibbons
Department of Liberal Studies in Science
University of Manchester

This essay is an extended reflection on the notion of objectivity as it refers to the systematic evaluation of the social consequences of technology, that is, to technology assessment. Technology assessment is both a movement and a set of techniques. In so far as it has its origins in the growing awareness among the peoples of the developed nations that technology can have bad as well as good effects and in their determination to draw governments into its effective control, it might be called a movement—a movement aimed at social responsibility in technology. On the other hand, insofar as technology assessment seeks objective, value-free analysis of the effects of technology, it represents the latest in a long line of techniques which purport to extend the rigors of scientific method from the physical and biological sciences to the economic, political and social ones. Among the earliest of these techniques, one could mention, for example, operations research, a problem-solving technique developed for military operations in World War II. Operations research proceeds by breaking up a complex problem into manageable pieces which are solvable and then, by arranging the individual solutions in a certain order, is able to achieve a *satisfactory* solution to the original problem. Operations research, now a recognized academic discipline, was followed in the post-war period by technological forecasting, systems analysis, futures research, and, most recently, by technology assessment. One need not possess the gift of prophecy to predict that over the next generation each of these will become an integral part of university curricula and so provide academics with an increasing number of techniques to pass on to students anxious to change the world for the better.

These techniques have in common the goal of systematic, objective, value-free analysis of the problems to which they attend. Thus, technology assessment strives to achieve such an analysis of the consequences of technology. Of all the techniques mentioned above, it claims to have the widest scope, being concerned with nothing less than the social, political and economic consequences of technology.

To be sure, many of the problems which the developed nations are experiencing today have a technological component and it is highly laudable to try, as technology assessment does, to anticipate the consequences, both good and bad, which could result from the widespread application of a given technology.

But, while one can sympathize with the aim of technology assessment, one cannot but be concerned at the trend its methodology is taking. As pointed out above, technology assessment emerged, and continues still, as a popular movement of concern for the impact of technology on society, but it seems to me that if technology assessment turns its attention to developing objective, value-free methodologies, it might well alienate itself from its social origins and develop into a technical discipline unintelligible to the very people who first voiced concern. Thus, the main theme of this paper will explore the notion of objectivity as conceived by many of the proponents of rigorous methodology as a precondition for technology assessment. We shall begin by discussing what objectivity is commonly understood to be by scientists and engineers since it is they who are in the vanguard of the movement for rigorous methodology. Secondly, we shall try to explore the notion of the 'good of order' so that we may have some grasp of what is involved in a determination of the social consequences of technology. Thirdly, we shall attempt to show that the facts of social organization demand an enhanced participation in decision-making. Finally, we note that knowledge about the social consequences of technology so obtained is more 'objective' than that conceived by the methodologists.

1. The Notion of Objectivity

The notion of objectivity is a complex one and, as it is used by scientists and engineers, it frequently involves three aspects. First, an analysis will be objective if it can, somehow, stand aside from the phenomenon to be observed. Thus, the scientist is believed to study and report on the "already out there now real." This form of extroverted objectivity has recently been criticized by Roszak.[1] Second, by stepping aside from reality, so to speak, to contemplate it, the scientist is able to free himself from making value judgments; hence the belief that scientific method is value-free. Third, analysis which does not attempt to proceed in this manner cannot claim to be either scientific or objective. A corollary of this position is that those who are functioning *within* the various systems that are operative in a society are behaving unscientifically and, some say, irrationally. From the point of view of the aims and intents of technology assessment, it is the tendency to separate the analysis of the social consequences of technology from the matrix of social decision-making in order to find objectivity that is most worrisome. It is worrisome because it appears to lack any awareness of the nature of social knowledge and of how such knowledge may be used to alter the courses of existing institutions. We will discuss some aspects of this situation after the brief introduction to the notion of the good of order which follows immediately.

2. The Notion of the Good of Order[2]

Technology assessment, then, is primarily concerned with evaluation of the social, economic, and political consequences of technology and, as we have seen, it has had its origin in profound popular concern for the continuing functions of the social order. An increasing number of individuals in society are mentally extrapolating the technological trends existing today and more often than not they come to the somewhat gloomy conclusion that either technology has to be more effectively controlled or society, as we know it, will disappear. Whether or not this judgment is valid, it expresses the conviction that technology is intimately involved with manifold social processes and that a too rapidly developing technology can, like a cancer, undermine the whole structure of society. It further implies the recognition that some form of social order is necessary if individuals are to achieve what they personally want out of life. Thus, while men persistently disagree about the 'right' form of order to be pursued, they would never intentionally opt for chaos. Order, then, is desired, more or less passionately, for what it can provide in terms of personal or group fulfillment. Because it is desired, we shall refer to it as the 'good of order' rather than the 'social order.' In this way, it seems to me, we keep in the foreground the notion that order is the outcome of the desires of men, operating in concrete situations and that all change, large or small, must be conceived and implemented through their ability to organize themselves in pursuit of the ends which they desire. Correlatively, we wish to push to the background the notion that scientific method, or indeed any technique, can short-circuit this route and so achieve the good of order by some other means. We could, then, rephrase the first sentence of this paragraph normatively and say that technology assessment, in trying to understand the social consequences of technology, *should be* trying to identify the main components that constitute the good of order and to examine the effects that new or existing technological programs might have on them. This norm can be made more concrete by considering the notion of the good of order a little more fully and by illustrating what is meant by a few examples.

It has been asserted that because no one wishes to live in total chaos, individuals in order to achieve their ends have organized themselves from the earliest times into families, groups, communities, societies; in other words they enter into different forms of cooperation. The concrete manner in which these various modes of cooperation actually work out is what is meant by the good of order. It is distinct from instances of the particular good—the individual goal or objective to be achieved—but it is not separate from them. The good of order regards the attainment of individual ends not singly and as related to the individual but *collectively and as recurrent*. Thus, for example, the ability to drive into the countryside on a particular day is, for me, an instance of a particular good. But roads to allow all members of the community to make the journey in safety is part of the good of order. Indeed, it is precisely the difficulty in maintaining this aspect of the good of order that has caused people to question whether or not the private family motor car is the best means to that end. Again, education for me is a particular good.

But education for everyone in society that wants it is another part of the good of order.

The good of order, however, is not merely a sustained succession of recurring instances of particular goods—the ability to satisfy recurrent human desires. Besides the multitude of individual goods that are achieved there is the order that sustains it. At the most general level, the good of order, then, consists basically in:

(i) the ordering of individual operations, skills, etc., so that they are genuine cooperations and, thus, ensure the recurrence of all effectively desired instances of particular good (e.g., education for every individual that desires it), and,

(ii) the interdependence of effective desires or decisions with the appropriate performance by cooperating individuals. Thus, a society which consistently ignores the wishes of one of its constituent groups invites the deterioration or, ultimately, the complete withdrawal of cooperation of that group; and because society is a complex web of such cooperations, breakdown in one area quickly spreads to another. Witness, for example, the various trade unions becoming aware of their mutual relationships. Nowadays, a strike of coal miners can paralyze the electricity generation industry—and not because of of the lack of available coal—but because grievances in one union rally the support of many.

The idea of the good of order has been summarized most succinctly by Lonergan:

It is to be insisted that the good of order is not some design for utopia, some theoretic idea, some set of ethical precepts, some code of laws, or some super institution. It is quite concrete. It is the actually functioning or malfunctioning set of "if-then" relationships guiding operators and coordinating operators. It is the ground whence recur or fail to recur whatever instances of the particular good are recurring or failing to recur. It has a basis in institutions but it is a product of much more, of all the skill and knowhow, all the industry and resourcefulness, all the ambition and fellow feeling of a whole people, adapting to each change of circumstances, meeting each new emergency, struggling against every tendency to disorder.[3]

An important corollary to this description of the good of order is that society functions, not by adhering to some master plan but on the basis of an incomplete set of insights—often referred to as practical common sense. This common sense knowledge does not reside in the heads of one or a few individuals, nor indeed, in the computerized memory of some data bank. Rather, it is dispersed throughout the community. From this it further follows that those who would change society or protect it from perceived dangers must either possess a master plan (which we doubt) or be willing to find out how society actually functions by discussing the situation with those in the know.

With these illustrations of the good of order—and its complexity—it should be becoming clear that the effects of technology on the manifold social processes are operative will be multiple and often indirect. Further, if the "facts" of the social situation are, as we have indicated, widely dispersed through the community, how, then, does technology assessment propose to go about achieving its goal of investigating the social consequences of technology and what claims do its methods have to objectivity? It is at this juncture that the ideals and concrete possibilities of technology assessment stand in sharpest contrast. Much of the discrepancy, we feel, can be traced to a narrow view of social processes and the method of achieving objectivity in this context.

The proponents of technology assessment do not seem particularly interested in elucidating the manifold of "if-then" relationships guiding operators and conducting operations which allow the good of order to be maintained and *then* raising questions about how existing or future technological developments may be expected to effect them. Much of the recent methodological development appears to be beginning from an opposite point of view; by speculating on the future internal development of technology and *then* by speculating on the likely consequences of this likely development for society. Technological development is considered as an *independent variable* and isolated from the concrete social process which is making it, or will make it, a concrete possibility. The method of achieving an indication of the likely social consequences in this context is usually some variant of a technique called *cross-impact analysis*. In this, a round table of experts is invited to speculate on the likely social consequences of a given technological development; and while it may be argued that there is no better way in which to proceed, still, one wonders how such a procedure, in spite of elaborate computer programs, can claim to be objective. The methodology aims at preparing some comments on the social consequences of a given technological development, as a package, so to speak, which can then be presented to policy-makers as an aid to them in reaching decisions about future technological developments. Because such analysis has failed to *involve* those charged with operating the manifold "if-then" relationships which exist, it is almost certain to lack credibility and consequently be of little effect in determining the outcome of the actual decision-making processes. Thus, it seems to be a real possibility that by pursuing the sort of detached, objectivity which we referred to above, technology assessment may well be rendering itself impotent to alter the course of events. From this, however, it does not follow that technology assessment will not continue to flourish and eventually become a recognized academic discipline!

3.   The Technology Assessment System

If we are skeptical about the claims to objectivity made by those who would employ the procedures described in the previous section, it does not follow that there is no way to investigate the social consequences of technology. The notion of the good of order which we have put forward has stressed the importance of understanding the manifold "if-then" relationships which are operative. We have further suggested that such understanding of these relationships as exists resides in the minds of the individuals who are charged with their operation; there is no theoretically formed master plan available which outlines the various operating social processes and their interlinkages. Consequently, the more effective control of the social consequences of technology lies in the direction of increased participation in the decision-making processes concerned with technological developments. Before attempting to show how this might be achieved it will be necessary to introduce and discuss the concept of the technology assessment system.[4]

The technology assessment system comprises those individuals and groups who are or should be concerned with the development of a given technological capability. The individuals and groups, let us call them actors, are bound together either formally or informally by a mutual interest in one or other aspect of the development of the technological capability. It follows, then, that one may expect the composition of actors in the technology assessment system to vary with the technology under consideration. Therefore, if one wishes to come to an understanding of the social consequences of a given technology one must begin by bringing together the actors involved. After all, it is they who possess the knowledge of the situation—there is no other. Even extensive interviews with the individual actors would not convey the depth of knowledge required to effectively alter or maintain the course of events. In any case, it seems wasteful; why not make *use* of the actors and the knowledge they have? The obvious difficulty is that each actor is concerned with a slightly different aspect of the good of order and consequently perceives the situation differently. Thus, in drawing together all the actors who are concerned with a technological development, one is increasing their mutual awareness not in the abstract but concretely. A further aspect of the technology assessment system concerns the identification of those actors who *should be* but are not involved in the decision-making processes about a given technological development. One of the novel elements of technology assessment is manifested in its concern for minority groups, which, for one reason or another, are unable to organize themselves in such a way as to have their points of view articulated and seriously discussed. The technology assessment system would be incomplete without those "should be" actors.

Still, the drawing together of the various actors in the technology assessment system will not occur spontaneously; narrowness of perspective, institutional barriers, and the heavy demands made on individuals by their organizations all operate to keep knowledge compartmentalized and allow decisions to be made which may adversely affect other groups or even society at large. So, there is a need for some sort of body to help to *coordinate* the actors and *interpret* the diverse perspectives in a language which is understandable not only to all the other actors but to the public at large. The principal advantage of this procedure—one would hardly call it a methodology—is that it keeps the informational and decision-making aspects of the policy process in close contact throughout the search for the possible social consequences of a technical development. In this way individual perspectives will evolve in a fuller awareness of those of the other actors and, possibly, these new insights will lead to more comprehensive socially-oriented policy decisions. This paper is, perhaps, not the place to go into the 'nuts and bolts' of the coordinating body. Suffice it to say that its objective and mode of functioning would be very different from the Office of Technology Assessment recently established to aid Congress in reaching decisions about new technological programs.

4. Objectivity in Technology Assessment

This essay began by considering the notion of objectivity in the context of technology assessment and it may not yet be clear how the procedures we have described for articulating the technology assessment system relate to that notion. To be sure, the procedures which we have outlined cannot claim to be objective in the sense of detaching oneself from the social process in order to examine it; precisely the converse in fact. We are beginning from the position that social decision-making—those decisions concerned with pursuing the good of order—are subjective and value-laden but we do not discount them as irrational. It is simply a fact that knowledge—the common sense knowledge through which our society functions—is dispersed throughout the community and that it is based upon an incomplete set of insights which must continually be improved, corrected, revised, and perhaps discarded. Further, we are assuming it is more sensible to begin from what is, rather than from what one would like to be. Consequently, the notion of objectivity we have been heading for is that which results from a cumulative build up of judgments of fact and of value. Now, judgments of fact and value can be error but provided one is open to different perspectives, to new information and to a spirit of cooperation, false judgments will be corrected as new insights come to light. Louis Wirth, in summarizing Mannheim's idea of the nature of social knowledge expresses succinctly our point of view,

> Hence insight may be regarded as the core of social knowledge. It is arrived at by being on the inside of the phenomena to be observed . . . It is the participation in an activity that generates interest, purpose, point of view, value, meaning and intelligibility as well as bias.[5]

It seems to me that the articulation of the technology assessment system will help to create the milieu in which the required insights will occur and, if this is so, the set of judgments which follow will, if continually open to correction, revision, and adaptation, gradually give rise to authentic objectivity concerning the social consequences of technology.

E. H. Carr has observed that no science deserves the name until it has acquired sufficient humility not to consider itself omnipotent, and to distinguish the analysis of what is from the aspiration of what might be. Technology assessment took its rise from widespread observations of the deleterious effects of uncontrolled technological expansion and the overwhelming purpose which is dominating and inspiring the pioneers of the new techniques of technology assessment is, obviously, to prevent recurrences of the same thing. Like other infant sciences, technology assessment has been markedly and frankly utopian. "It has been in the initial stage in which wishing prevails over thinking, generalization over observation, and in which little attempt is made at a critical analysis of existing facts or available means."[6] If technology assessment is to mature as an intellectual discipline it surely must make this transition from a utopian ideal to a critical one.

REFERENCES

1. T. Roszak, *The Making of a Counter Culture* (New York: Anchor Books, 1969).
2. B. Lonergan, *Method in Theology* (London: Darton, Longman and Todd, 1972). Much of the discussion on the notion of the good of order has been drawn from pages 47-55.
3. *Ibid.,* p. 49
4. The concept of the Technology Assessment System is discussed in detail and illustrated using a case study in a forthcoming report of the Science Council of Canada entitled *The Tehcnology Assessment System.* Background Study No. 30 by M. Gibbons and R. D. Voyer.
5. L. Wirth, Preface to Karl Mannheim, *Ideology and Utopia* (New York: Harcourt, Brace, 1949), xxii.
6. E. H. Carr, *The Twenty Years' Crisis (1919-1939)* (Harper Torchbooks, 1964) p. 8.

James K. Feibleman's View of Technology and the Problem of Ideologies

Stanley C. Feldman
Attorney
Washington, D.C.

Conceived most broadly, the inquiry into the interface between technology and the state is an inquiry into the most profound and critical of problems in contemporary civilization. For such a task it is especially appropriate to consider the comprehensive views of the eminent American philosopher, James K. Feibleman. Feibleman's all embracing conception of philosophy finds an underlying unity in whatever problem we may happen to be considering, and the present is no exception.

For the immediate topic, the unifying idea is that moral considerations are very much involved in the making, selection, and use of artifacts.

The making and use of artifacts is very much a part of all human endeavor, and so what is to be made and used is primarily a moral question. Technology is a subdivision of ethics. [1]

Thus, as Feibleman sees it, *all* human activity is answerable to ethical theory—thoughts and feelings as well as actions. And it is the state, for Feibleman, that organizes human activity, establishes its concrete social morality and a government which is involved in its establishment and maintenance. [2] But there is something prior to the concrete social morality, and that is theoretical ethics. [3] Before the social morality of a society is established, it must first be decided what it ought to be. The state of the future, including the possibly perfect state, would therefore depend upon present speculations of theoretical ethics.

But if it is clear that the technology of a society defines what that society *can* do, who is to decide what it *ought* to do? In short, how is it decided in some prior fashion what direction politics should take in the field of speculative ethics? The fundamental nature of such a choice will certainly affect every activity of the society, and could determine what technology is to be encouraged and used and what is to be eliminated. Such grave moral considerations are involved that the choice of direction is not one that technology itself should be allowed to make.

While philosophy may contribute to the assessment of alternatives, it happens that other institutions supply the decision makers and provide direction.

An ethic is chosen before anyone is aware that a problem exists, and it is made by a few who are aware at the expense of the many who are not.[4]

For Feibleman, goal-setting is a first necessity in politics without which nothing can be done, and politics "is the theory of how to get done what according to ethics ought to be done."[5] Government, then carries out what the established morality dictates.

It is important to note here that Feibleman's concept of the state does not in most cases determine morality, but simply acts as custodian for the morality which is established.

The state is usually neutral with respect to values, and administers whatever values the society acting through its leading institution wishes to establish.[6]

Fundamental social change is the result of three public moralities: the outgoing, the established and the incoming.[7] Accordingly, the morality of a society constitutes a pool into which the government can dip in carrying on its work, as, for example, when it mkes new laws to cope with new circumstances. Only in the instance of the extreme disorder of social revolution or war may a stated morality be overthrown and the establishment of another be sought.

The oriental proverb, "Peace is the dream of the wise, war is the history of man," seems especially true today. Preventing war is apparently *only* a dream. Feibleman defines war broadly as "the extension of conflict into extreme violence between peoples organized into political units,"[8] and as "the planned and organized collective aggression of one people against another."[9] While both definitions are consistent, the second is analytic and it is this one that leads him to the following conclusions:

The inevitability of war is a characterization that has seldom been made. I have tried to show that it is inherent in the aggressive human animal who turns his unfulfilled need over to the more powerful state. But I submit that the causes of war, in addition, remain largely unknown.[10]

The fact that the inclination if not the actual manifestation toward war exists in all countries appears to confirm the view that extreme destructive aggression is natural to the human condition. But in the end, as Feibleman acknowledges, the causes of war are essentially unknown even though the biosocial explanation is basically sound. One known cause of war Feibleman does see that may not be reducible to a biosocial fact is "the pretension to the possession of the absolute truth."[11]

If a state thinks it has [possession of the absolute truth], there is no choice left but to seek to impose it no others—always, of course, for their own good.

Given two or more such states, war is inevitable; even with one, there could be organized military resistance. For the absolute truth may be one of political economy—the Soviet Union—or one of religion—Islam—but it does not matter since the result will be the same. According to Muslim law, for example, the world is divided into a domain of Islam and a domain of war. Had the entire world embraced Islam, we are told, this second domain would not have been necessary, but under the circumstances of resistancce to conversion it is.[12]

Although Feibleman sees too much ferocity inherent in man's nature at the present for an eventual resolution of recurrent wars, he does have a long-range program requiring a global super-state which may be the only means for achieving an international peace.[13] The subject of the creation of a new political system to eradicate war is not, however, within the limits of our present discussion.

Feibleman argues that before society in the round can avoid the hazards of societal instability, it must understand the connection between ideology and exigency as an ultimate one. If an ideology is "a conception of the rights of man based on a theory of reality," then every ideology is in a sense a utopia.[14] The problem of ideologies, he concludes, is simply that there is a plethora of utopias. Which one ought to be adopted?

That absolutism has over and over been the story of utopian ideologies points to a basic flaw. The very idea of an inflexible ideal contains a contradiction. As an ideal it sets the goal for action but as inflexible it cannot be a goal since actions never conform inflexibly. . . . Thus, ideologies have succeded to the extent that aiming at them has made actual governments possible; they have failed to the extent to which, in applying them, actual governments have operated in an uncompromising fashion. Clearly this means that somehow an ideology must be designed that will meet all the usual requirements and yet not be so inflexible as to produce havoc when applied. To that extent it would fail as an ideal.[15]

As a result of all the variables which are always involved in any social situation, and in view of the scarcity of reliable knowledge and the mistakes which he sees have so often occurred in the past, Feibleman finds a dilemma that he states as a political rule: "*since perfection is always absolute, calls for perfection are also calls for absolute action; but action is never perfect though aimed at absolutes.*"[16]

Feibleman suggests, therefore, that ideologies must be framed in limited and tentative terms mainly because of their half-exigent natures. "An open-ended and flexible ideology is what we may call, paradoxically, the ideological ideal."[17] For such a task, as is usually the case, the more we investigate the greater the number and the difficulty of problems we uncover. Given all our ignorance, however, we have been well advised by Professor Feibleman to see that the discovery of ideals is a speculative field which must be explored before better postulates are chosen for belief.

70

REFERENCES

1. *Moral Strategy, an Introduction to the Ethics of Confrontation* (The Hague: Martinus Nijhoff, 1967), p. 32.
2. *The Reach of Politics, a New Look at Government* (New York: Horizon Press, 1969), p. 77.
3. *Ibid.*
4. *Ibid.*
5. *Ibid.*, p. 58.
6. *Ibid.*, p. 57.
7. *Ibid.*, p. 58.
8. *Ibid.*, p. 257.
9. *Ibid.*, p. 264.
10. *Ibid.*, p. 268.
11. *Ibid.*, p. 269.
12. *Ibid.*
13. *Ibid.*, p. 275.
14. *Ibid.*, p. 95.
15. *Ibid.*, p. 96.
16. *Ibid.*, p. 97.
17. *Ibid.*

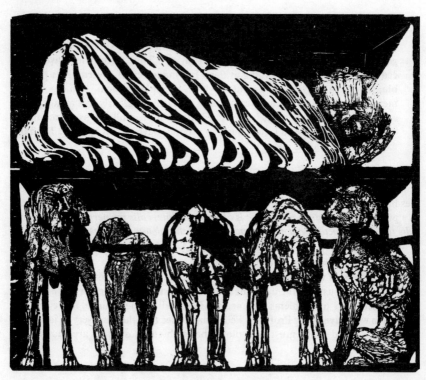

"Sleep", 32½ x 39½" (Woodcut).

Using Technology to Change Society: A Political Problem

Brewster C. Denny
Dean
Graduate School of Public Affairs
University of Washington

More than 20 years ago, I went to M.I.T. to teach in the School of Humanities there. My training was in social science and history. My skills were verbal, and my analytical approaches were very far from what would be called rigorous by today's social science standards. I was one of several young faculty members at M.I.T. variously trained in history, literature, political science, and economics. Our mission was to "humanize" the engineer. This mission was based upon the assumption that science and technology were greatly altering society in the years of wonder and mystery then unfolding and would likely profoundly alter life on this globe and in this universe in ways then unforeseen or at the most only dimly so. The assumption was that the scientist and engineer who would invent a new future for man must be tempered with knowledge of history and politics and philosophy. The humanities were thus seen as an input, a tool or a dimension for those men who would not only change the technological tools and implements of society but, because these would be the instruments of that change, would also make policy — public policy. With the immense technology created by that post-World War II generation of young men, there opened for mankind opportunities of almost unlimited good and dangers and hazards so great as to threaten the survival of life on this planet itself.

The "humanizing" of the scientist and engineer is and was a noble cause. Whether it is more the product of a conscious educational program and humanities requirement, or, so I believe, more the result of the innate "humanness," intellectual curiosity, and public consciousness of a great generation of scientists and engineers, it is clear that the engineer has been "humanized." The question here is whether educational humanization of the engineer should be the principal mode in which we approach the intellectual and educational problem of making technology work best for high public and social purposes and for inoculating the

72

world against the disasters with which technology threatens us. I think not. The purpose of this paper is to say why and to propose an alternative.

Technology makes possible both goods and evils of a scope never possible before in human history. Technology's first achievements were in the struggle of man against the hostilities of the earth's environment. Man tamed nature and put it to his use through technology. But as man enters the last third of the 20th century, nature is tamed. Man has the technological capacity to eliminate disease, poverty, and hunger. Biological knowledge now permits, or will shortly permit, fantastic interventions in the development of the human organism itself. Yet, overpopulation, environmental pollution, and modern weapons technology quite literally threaten and, if unchecked, promise surely the extinction of man. What man prevented nature from doing, man can now do. The same technological revolutions which have opened a wide prospect threaten the extinction of the species. Invention of the capacity to do great good has brought also the capacity to do great harm. Atomic energy and the internal combustion engine are easy examples.

What priority tasks does this general circumstance lay before us? Two examples are illustrative: First, the limitation of population growth to a zero rate, or even lower for a while, is essential. Second, there must be a major energy shift, particularly in advanced industrial societies, to those energy-producing technologies which neither pollute the environment nor deplete energy sources. These two examples by no means exhaust the possibilities but serve to illustrate the central theme of this paper.

Population control and a shift to a renewable or inexhaustible and largely non-polluting energy technology are both technologically achievable. The technologies to achieve them have been, are being, or should shortly be invented and will be further perfected by brain power now alive. And, to anticipate my theme, regardless of whether or not the scientists and engineers making those inventions are socially aware, humanized, or trained in and knowledgeable about public policy matters, the responsibility and burden of setting the priorities for these tasks and putting these technologies to work will belong to a different profession than that which developed them. Their work will be harder and, because time is urgent, even more important. The introduction of the technologies thus developed into the life of this planet to the degree necessary to avoid the disaster of overpopulation, environmental pollution, and exhaustion of energy sources is a political and not a technological problem. It is the responsibility of political representatives, many of whom should be scientists and engineers by training and experience.

I don't mean to oversimplify. The technologies in both of these areas are far from perfected and much in their technical development must bear in mind the problems of introducing the technologies into the public sector. Birth control methods that have no harmful side effects, that are simply administered, that can be administered infrequently, and which lend themselves both to individual decision and to methods of public intervention are not quite yet here. The technology of the slurry reactor and the technology to make such a reactor foolproof even against earthquakes, sabotage, and human error are not fully achieved. But the technologies in both of these vital fields in terms of difficulty of achievement fade to virtual insignificance by comparison to the public policy and political problems of effectively introducing them. It must be specialists at *that* business — whether they were once doctors, engineers, or public health administrators or not — who perform the critical tasks. And *that* business is politics.

The mix between the technologies which must be invented or perfected and the political problem of introducing them effectively into society is not always as simple as the examples given above. Man is becoming more than dimly aware that public intervention in the rearing of children and the provision of intellectual, emotional, and physical nutrition is as essential to the solution of the terrible problems of poverty and racism as programs of economics, welfare, and job opportunities are to this vital cause. Affluent America now knows that its society will not survive without winning the war on poverty and racism and an immediate reversal of the present trend towards disintegration of primary and secondary education. Here the knowledge base is far less developed. Rapid development of knowledge about the emotional and educational nutrition of children and its relationship to the physical nutrition of the poor and the disadvantaged plus massive public intervention in pre-primary education and the public schools themselves, as well as in the economic status of families below the poverty line, may be the essential inoculation against disaster which will be required. In these instances, the knowledge problem — that is, the technology — is far from solved. Worse yet, the political and social problems of public intervention are far more serious than even telling Catholics they can't have children.

Modern technology lays before us tasks of disaster inoculation and of the full utilization of the advantages and promise of science. Technology makes both possible and necessary new kinds of social intervention, new kinds of intervention in behalf of the public interest into the private sector — not just the private sector of business now so largely riddled by forms of government intervention in the name of the public interest as to render that type of intervention somewhat understandable. It becomes a public intervention into private and personal life, into man's castle, his bedroom, and even his automobile and his segregated trade union or country club.

Interventions of this sort are, of course, not unknown. The Constitution of the United States provides both authority for and sets limits upon governmental intervention into personal life. The technology of epidemiology and immunology made possible a kind of intervention which is now quite well accepted and which is clearly constitutional within the meaning of the police power. In these instances, the power of the state is used to prevent the spread of infectious disease. Water supply control and vaccination are designed to prevent one individual from spreading his disease to others. Disaster or emergency threats by plagues and epidemics — known phenomena from early recorded history — are accepted as sufficient threats to justify massive public intervention in private life. Not all these kinds of interventions through the public health power are fully accepted: witness fluoridation, smoking, alcohol, and drug control.

In the field of national security, the power of the state to intervene in private lives through the imposition of martial law, the draft, and all the other measures associated with the need for the state to defend the state itself and thus its citizens is an area of well-defined, if currently troubled, social intervention. The public and Constitutional doctrine have accepted a degree of public intervention in the economy largely designed to prevent the disastrous affects of the cyclical fluctuations of the economy, particularly those caused by unemployment, although the knowledge base is still limited. The actions of the 1930's and the Full Employment Act of 1946, as well as a range of government interventions in the economy beginning in the last century with the Sherman Anti-Trust Act, were recognitions of the need to give government authority to inoculate the economy against economic disaster even if the means of inoculation involved considerable intervention into private affairs. In the present economic crisis (1970), the principal criticism of the President by business is that he is intervening too little or in the wrong ways in the economy. Thirty-five years ago in another economic crisis, federal intervention itself was criticized as evil and unconstitutional.

In the three analogies above — public health, national defense, and the economy — in which public intervention in private affairs is accepted and understood within a set of Constitutional ground rules constantly tested, there is a common element. In each there is a history of known disasters of such magnitude that massive inoculation against disaster was politically supported. Thus, the plagues, the crushing of unprepared nations by tyrants, and the terrible human disasters of great economic depressions were part of the experience which led the body politic to accept public intervention. Even the city manager movement — a valiant, but only partially successful, effort at inoculation against the terrible weaknesses of local and state government in this country — gained public acceptance in important measure because of rampant and scandalous corruption in city government.

The imminence of disasters we have never had is harder to use as a basis for public intervention than the sure knowledge of one which man has experienced. Intervention in personal life to prevent an epidemic of smallpox has become politically acceptable because mankind has experienced such epidemics. Birth control to prevent a biological and nutritional disaster never experienced and not easily imagined is another matter.

In seeking public and educational strategies to meet the challenges of both the opportunities and threats of technology, it is essential to avoid two dangerous oversimplifications of the problem. The first is the danger that the interdependence of the public and private sectors and the blurring of the lines between them, which is implicit in what we have discussed above, might lead to the belief that it is no longer important to keep the public and private sectors distinct. The argument that the line between the public sector and private sector has been blurred almost to indistinction is easily supported. Is the aerospace industry really private business? Does a universal system of medical care sponsored by the Federal Government really permit the practice of private medicine? Do the complex national tax systems now used actually permit the free internal operation of business? Does increasing public control of environmental matters really leave major business decisions to market considerations? This familiar litany of examples of the blurring of the lines between the private and public sectors has led many to despair and a few to rejoice in a premature assumption that American society has, after all, become socialistic. It has led others to believe that there is no longer a distinction between public and private man and that the education, ethics, and perspective of each are so much the same that public servants should be trained in business schools and that business management and public management are the same thing.

But is this blurring of the line between the sectors an illusion? If the trend — largely due to technology — is for more public intervention on behalf of the public good and in the name of the public interest than in the past, then is it not more important that the public interest itself be better understood and distinctly identified? Is it not more important that public persons behave as public persons, and the distinction between a man's public and private acts be understood and accepted by the society? Is it not more important that public men be reared and trained in the tradition and principles of a separately identified public ethic rather than the business management ethic of private sector? The answer to the conflict of interest issues raised by the increasing number of people who move in and out of public life is not to declare that the distinction no longer matters, since there are so many of them, but to insist on clearer ground rules for such behavior. The disasters which nearly overtook the American political system in the early days of industrialization can be attributed in part to the fact that men like Roscoe Conkling and James G. Blaine neither recognized nor honored the distinction between public and private when the blurring phenomenon first occurred over the new technology of railroads.

A second and closely related threat lies in the new decision-making technologies — such as systems analysis, PERT and PPB — which engineers and economists, for the most part, have developed. The popularity of these technologies rests principally on the as yet unproved, but widely accepted, notion that decision-making techniques made Robert McNamara an enormously successful Secretary of Defense and on the spectacular achievements of management technologies in the space program and similar publicly financed, but privately executed, engineering ventures of great complexity and magnitude. The danger does not, of course, lie in the management of specific programs and projects through the use of highly sophisticated techniques, whether in the public or the private sector, nor in the use of such techniques to pose alternatives and order information for decision makers. The danger lies in the realm of public policy and public decision-making itself and the ways in which project and program management systems and executive information systems come to be viewed as decision-making technologies and, all noble protests aside, decisions themselves. These combined with the capacities of public opinion analysis present a severe threat. Public opinion becomes an "input"; decision-making, a black box run by experts; and public policy, an "output." In such a model, the government is only responsible to the electorate in a broad and general sense, and the complexities of government decision-making are left to experts.

No amount of expertise and decision-making technology can substitute for real participation. Policy makers must not be responsible to the public only "in the end." That's no responsibility at all. They must be responsible in the beginning and the middle, for the take-offs and the flights, as well as the landings. If the public is only an "input" and not *in* the black box, there is no self-government. This is a major part of the message that millions of young people are trying to send to the establishment these days. They are not inputs, throughputs, or outputs. They want inside the black box and want it opened up. A few, of course, want to take it over and close others out.

Government is not an industrial engineering problem, a systems analysis problem, a business management problem, a market analysis problem. Nor is it a conglomerate of all the skills and techniques which have been designed to deal with such problems carefully packaged and modified to operate in a slightly different environment. The management specialist, wherever he comes from, often makes it sound very easy. It's really the same problem, he says, with a few little differences that, of course, we'll leave to the political process. We're just talking about administration, he says, not policy, as if they were separate things, which they are not. There is great danger that man's pathetic desire to put decision and responsibility in a mechanism, in a process, or a machine may at last have found, in the period of big government and big science, methodologies and technologies of decision-making which will take man off the hook for good. If decision-making is left only to specialists who can master a complex technical decision-making apparatus, self-government is by definition dead. It is absolutely essential that the political system designed to deal with the problem of massive, complex, and difficult intervention in society must be ventilated by publicness and not closed by lingo, experts, systems analysis, the "administrative sciences," and a whole range of techniques which can help the process but must never substitute for it. If engineers have invented decision-making machines, they have invented the end of self-government.

These warnings are reflections of a central point. The greatest challenge of the potentialities and dangers of modern technology are to the political process itself, both in the sense of the survival of the fragile and noble idea of self-government and in the capacity of the political process in the public interest to facilitate and control the introduction of technology in new ways and for new purposes into society. The highest priorities lie in the fundamentals of that political process itself — terribly underdeveloped compared to technology. This underdevelopment, however, is not technological as so many believe.

It has become commonplace these days to ask the educational community to meet many of the urgent requirements of society. In many cases, the educational community is the wrong place to ask for magic and answers. The immense political tasks which lie ahead of inoculating against the disasters of technology and fully utilizing the great opportunities which technology lays before mankind are by no means solely problems for universities and intellectuals. For these questions are no less than the health of the body politic itself. And that is everybody's business. But the educational chores and educational priorities are deeply involved in these tasks. As I see it, there are four high-priority tasks for higher education in the broad business of enabling universities to participate effectively in making technology serve society and in assuring the survival of man and free institutions including universities:

1. The development of a more alert, better educated, more publicly-oriented citizenry to demand, accept, and control the necessary public interventions and to operate, participate in, and keep honest and responsive the process of government itself.

2. The preparation of the political leadership and public servants who will be required to meet these immense responsibilities.

3. Academic research and curricular development on government policy itself, the evaluation of public policy and programs, the improvement of public institutions, particularly local and state government and the institutions of education themselves. This means a very major shift in the emphasis and scale of social science research.

4. The continued education of scientists and engineers, first to do their scientific and technological tasks, particularly those of highest priority and, second to continue to do them — as I believe they have in the past — with full regard to the public and human aspects of their tasks.

All of the above four tasks are immensely important. My own view is that the fourth is the one which is farthest along, which has already received the greatest proportion of higher education resources, and which needs the least new assistance and change. Inherent in all of the other three tasks are some fundamental questions long neglected to which scientists and engineers should pay attention. In fact, I would argue that the best thing that engineering educators can do is to demand, as many have, that the first three long-neglected tasks be now raised high in priorities.

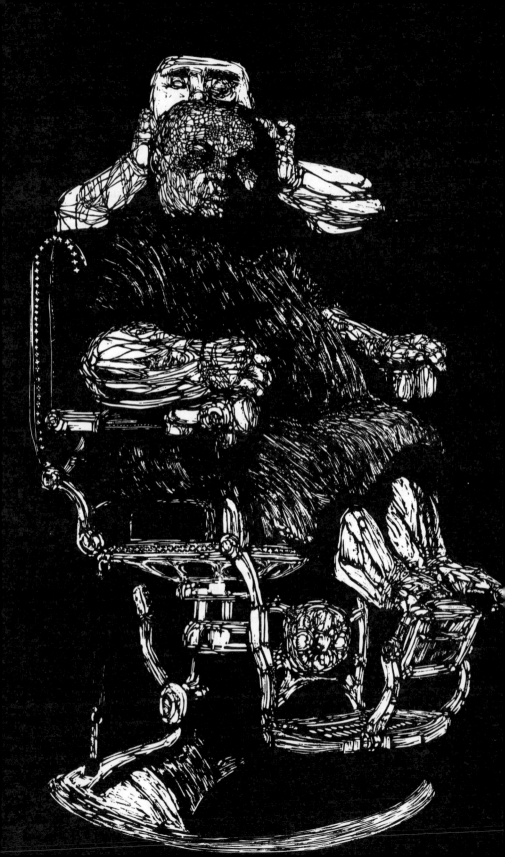

In all three tasks, the role of science and engineering is critical. I am not proposing that the education of the public, the education of political leaders and public administrators, institutional improvement and research on government institutions and government policies be left only to the social scientists. All three of those high priority tasks must be done with a very much more substantial input coming from the real world of science and technology. I, personally, believe it is more important that social scientists learn some real science and technology, particularly in the fields that they write about, such as environment, than that engineers learn political science. The technical education of social scientists and governmental generalists is far more an underdeveloped area than is the social science and political education of scientists and engineers.

All four of these tasks are closely allied. Each requires a conscious redesign of the educational system. Each requires a change in priorities in the educational system. Each is fundamental to the operation and survival of a democratic society in a time of great technological opportunity and potential technological disaster.

The highest priority would be for the education of the citizens. If democracy is to rest upon the masses and be anti-elite, the really key questions about democratic institutions concern the capacity of those masses to perform their central role in the operation of a democracy. So much of our public discussion of critical decisions in the technological age has rested upon the Presidency and the awesome decisions that he must make, that it is often overlooked that there is one other decision-maker in our society who has an even more awesome responsibility. That is the voter who selects the President. For, as Aristotle wisely said, "There are no more momentous duties than those of electing officers of State and holding them responsible." The priority task, then, is again in Aristotle's words, "the education of the citizens in the spirit of the polity." This is an immense task before education — perhaps even more important than teaching reading, which we aren't doing very well.

What will be, in the context of what we have said above about the political nature of the challenges of technological society, the posture of universities? Will we essentially continue to use incremental budgeting and simply shrink everybody's budget more or less equally and fairly, except that very able entrepreneurs working on very popular and scientifically exotic problems will get a little more, and those below average in political clout, glamor, and visibility may get a little less than average? I suspect that the above will roughly be the way resources are divided in this period of shrinking budgets. If that is the case, we can expect to see the continuation of some of the following trends:

Schools in this country designed to prepare professional public servants on an interdisciplinary basis will rock along at an annual production of such students at a rate of about 600 to 700 for the country as a whole — a number smaller than the total student body of many major engineering and business schools in the country, a number smaller than the need for such people each year in Los Angeles County alone. Business schools and industrial engineering programs will grow and continue to increase educational research programs designed to apply modern management methods from the private sector to the operation of government. The number of social scientists trained, as a whole, will be likely to increase with a continued increase in emphasis on the behavioral sciences. But the number of social and behavioral scientists and engineers and scientists trained on an inter-disciplinary basis for the application of their disciplinary expertise to the public sector will likely remain the same — that is, few and far between. The absolute numbers of economists and political scientists trained to deal with public policy problems will continue to decrease. (Let me say parenthetically that I am not at all sure that higher education in general and the social sciences in particular involved know how to spend a major infusion of funds in the social sciences if such an infusion were to come, which I doubt.) Grantsmanship in some of the new challenging programs being designed to meet the problems of society and to make universities relevant to their needs will place greater demands on that very small number of social scientists who are competent to work at this difficult kind of interdisciplinary activity and make little or no provision for the training of their successors. Undergraduate education will remain a stepchild in many institutions, and the rapid growth and proliferation of community colleges and new four-year colleges in the country will not be accompanied by a commitment of the major university research centers to the development of a strategy, teaching materials, trained professors, and a distinct view of those kinds of educational opportunities. In educational lobbying, I expect that the pattern established by the higher educa-tion lobby during the first Nixon Congress will continue, and it will be an effort to save old programs and extrapolate their further development rather than to develop a new strategy and new priorities.

I greatly hope the gloom I forecast above is not true. Let me suggest some specific measures that might be undertaken:

1. A very high priority of public investment in higher education research on pre-school education and primary and secondary education, and nurtur-ing is essential. We must learn about how man develops before we can in-vest in remedial programs we don't understand.

2. A very major increase in national resources for the education of public servants along the lines envisaged in Title IX of the Higher Education Act of 1968, which has been passed and signed into law but is still unfunded. A 20 million dollar annual investment by the federal government could triple the flow of able, well-trained professionals into public service in this coun-try. Such training must include full education of such people in the realities of technology.

3. Closely related to these and to the needs of an educated citizenry is the need for the engineering profession to develop ways to educate non-scientists and non-engineers with sufficient knowledge of technology — real technical knowledge — so that they can perform their public roles and their citizen roles in the light of full appreciation of technology. How much better

it would be, for example, if the public already understood that the place to build nuclear reactors for electric energy-producing purposes may be in and near the city rather than in the country. In this connection, I urge vigorous support and expansion of the NAE's Engineering Concepts Curriculum Project into the college curriculum, and I urge that social scientists with knowledge of the public sector be consulted on the public policy aspects of the technologies involved.

4. I have mentioned above the need for education of the citizenry. This is an extraordinarily complex issue and deserves the direct attention of the public. Education for citizenship has faded, and the current undergraduate curriculum does not seem to be well adapted to that purpose. The demand of students for relevance would appear to present a tremendous opportunity. Thus, a high priority is a look at the university itself.

5. Development of centers of excellence for strengthening governmental institutions, particularly local and state governmental institutions, and particularly at the business of introducing technology into society is desperately needed. The new technology is calling upon all levels of government, but most desperately on levels immediately below national government (that is, including regional administration of federal programs) for levels of expertise and competence at program development administration which are simply not there. Worse yet, the mechanisms of citizen participation in the policy process itself at those levels are weak, not so much because they are badly designed, but because they have been so long unused. The need is less for new gimmicks like the ombudsman and "participatory democracy" and more for the strengthening and improvement of the existing institutions. In this connection, note that while modern communication technology makes possible a kind of public participation in the electoral process which may make the ideal of Greek democracy at last feasible, we are still fooling around with the "fairness doctrine," candidates "packaged" by ad agencies, and the financing of political campaigns by private financial interest and lobbies. Universities need to invent institutional devices to meet this challenge, and there must be an allocation of the ever-scarcer educational resources to this chore. This allocation can be more effective if the breathless proliferation of incompletely conceived, federally sponsored programs can be brought under control. Fifty water resources research institutes, a large number of centers of excellence in environmental quality, plus a sea grant program, plus the RANN program, plus the NSF State Science program, plus a wide range of special urban programs just don't make sense. They should be brought together into regional and national centers of excellence at these vital tasks. The universities are guilty of pork barrel politics in permitting this to happen.

The approach I have proposed above is, in a way, an appeal from the underdeveloped part of the educational establishment to the developed part to urge upon them a call for development, a call for an improvement of the social sciences, a call for an improvement of the intellectual institutions by which we collectively make our contribution to society. As David Truman has so effectively pointed out, a very large part of the energy in the social sciences in recent years has been towards the internal methodology of the disciplines. And this was needed and still is, for they are weak in many important respects. But a critical problem now is to bring the social science disciplines into a genuine participation in policy problems in ways that preserve and enhance the highest academic values and maximize contribution to identifying and solving problems and achieving opportunities which a lushly financed scientific and technological establishment has set before us.

The tasks for engineering educators in this process are immense. They must avoid the temptation of offering decision-making technologies to do the citizen's and the public servant's job. They must assist in clarifying what is public and what is private and insist on knowing the difference. They must move aggressively in the most underdeveloped part of their profession — the devising of ways to educate non-scientists and non-engineers in the technological and scientific realities of the time in which they live and the times in which we will live — or perish. And finally, in doing his or her own thing, the engineer is a citizen, too, and must continue to contribute to the development of priorities for the introduction of science and technology into society for inoculation against disaster and achievement of the high hopes which the engineer has so largely made possible.

Some Tasks
for
'The Humanities'

Max Black
Program in Science, Technology and Society
Cornell University

1. *In search of defining criteria.* Considered as a general term, the pretentious and now somewhat shopworn label of "the Humanities" has the defect of lacking any clear criteria for membership. Indeed, the traditional label, for all its self-congratulatory connotations, has now degenerated into a mere umbrella-word for whatever cannot conveniently be assigned to "Science" or the "Fine Arts." It may still be useful, however, to search for some criteria that serve to distinguish "the Humanities" from other studies and disciplines.

There can, I think, be no hope of finding an Aristotelian definition *per genus et differentiam:* the various humanistic studies are too variegated and share too many features in common with the natural and social sciences to admit of neat classification. But we can still hope for glimpses of some unifying conception, some good reasons for a contrast with Science and Technology.

I shall use "humanistic studies" throughout to refer to second-order activities of "placing", interpreting, criticizing and evaluating primary "humanistic" texts, works of art, and related artifacts, institutions and practices. Correspondingly, I count, for the purpose of this discussion, a "humanist" as a student, teacher, or investigator working at this "second-order" level — a literary critic, not a poet; a musicologist, not a composer; and so on.

2. *The concept of "the Humanities" is an "essentially contested" one.* I borrow the useful notion of an "essentially contested concept" from Professor W. B. Gallie, who applies it to any term, such as 'Art,' or 'Democracy' whose users, by adhering to competing traditions, contest the right to possession of the honorific appellation. Such contests for a linguistic emblem, settled by the triumph of one sect over its enemies, are characteristically marked by the proclamation of "persuasive definitions": *"True* Christianity is . . .", *"Genuine democracy is . . .",* *"Real* freedom consists in . . ." and the like.

Humanitas, it seems, was Cicero and Varro's latinization of the Greek *paideia,* and alluded for many centuries to this historical affiliation. For Roman theorists it steadily connoted the syndrome of features, the "human essence," that elevated Man above the condition of mere animality and gave him his proper rank above the beasts, if below the gods. This program has always been controversial. It proved hard to reconcile the "personal" and the "public," the aspirations of an individual with those of his social groups.

Man, in a conception reaching back to Aristotle, and shared by anti-humanists no less than by humanists, could become most "fully man" by not being an "idiot" — in short by becoming fully a citizen; and so the "good arts" have always been conceived as "serving ends beyond themselves — ends that involved the moral nature of man or his practical activities as a citizen or public servant" (R. S. Crane, *The Idea of the Humanities,* 1967). Given the restrictions upon citizenship in all societies in which "humanistic education" has been an effective ideal, it is hardly surprising that, for all its pretensions to universal validity, such education should always have been both elitist and vocational, serving in fact, whether in Greece, Rome, Renaissance Italy or Victorian England, the special interests of a governing class or their clerks. This is one respectable ground for current suspicion of "humanistic education": it is hard to advocate in good faith a vocational curriculum, designed for gentlemen and their literate aides, when the very concept of a gentleman has become an anachronism.

Tension between the "personal" and the "social" aspects of classical humanism has been exacerbated by background philosophical disputes as to how to define the "human essence," and by endemic wrangles between curriculum makers. No wonder, then, that no agreement has long prevailed concerning the content of a humanistic curriculum.

The moral to be drawn from this Cook's Tour of the history of humanistic studies is that the concept here proposed for investigation is problematic in the extreme. Our choice is not between relatively well-articulated sub-traditions: contrary to popular belief, there never has been a Golden Age of humanistic education. The task is not so much to "save" the Humanities as to *create* them.

3. *The underlying presuppositions of earlier humanistic programs need substantial revision.* The background assumptions of the powerfully persuasive conception that there ought to be a distinctively "humane" education, designed to elicit those powers and virtues that distinguish men from brute beasts, can be captured in the following, deliberately simplified contentions.

(A) *Men are distinct from and potentially superior to other animals.* The "dignity of man" consists in his capacity, through the exercise of deliberation, reason, and choice to realize this potential superiority.

(B) *There is a human essence or "nature," present in all human beings, the same in all of them, that sharply distinguishes them from mere animals.* Thus the difference between a man and a brute is like that between silver and iron, one of "kind" not of "degree," innate, apt for full realization by suitable education and training, but impossible to alter.

(C) *The human essence is essentially good.* Classical humanisms, to the degree that their aims have been secular, have been guided by an optimistic rejection of original sin. The good for man is to become "fully human," by the full realization of the essential humanity implanted within him; in so doing, he becomes necessarily a *good* man and so, also, a good citizen. Humanistic education, thus conceived, necessarily has moral and social implications.

(D) *There are distinctive "arts" or "disciplines" peculiarly suitable for aiding in the realization of essential "humanity."* It is a noteworthy basis of continuity with present-day conceptions that linguistic studies — "grammar" and "rhetoric" — should always have been recognized to have a central humanizing importance. For if anything prominently distinguishes men from brutes, it is in their power to communicate by speech as members of a community possessing a common language.

(E) *The distinctively humanistic disciplines must include some that are non-scientific.* Science is not to be excluded from "the Humanities," as earlier humanistic curricula amply testify. Yet science has only a limited and partial educational value because, in its search for "objectivity," it is committed to "suspending" certain distinctively human interests and preoccupations.

(F) *The overall aim of humanistic education may be summarily described as the development of Reason.* Cf. Erasmus: "What is the proper nature of man? Surely it is to live the life of reason, for reason is the peculiar prerogative of man" (quoted from M.H. Abrams, "Humanism," in *A Glossary of Literary Terms, 1971).* If thesis (E) is accepted, the character of "reason" itself becomes problematic: are the same methods and procedures as suitable for literary criticism as for physics? Perhaps no question is more important for advocates of a distinctively humanistic education.

Taking now the position defined by the above six contentions as an "ideal type," and a plausible springboard for a contemporary program, let us consider how much of it remains tenable in the light of present-day knowledge. Only the fourth and fifth contentions can survive without serious modification. While agreeing that man is strikingly different from other animals, we must acknowledge continuities that discredit the oversimplified ancient conception of a unique, innate and sharply-defined human "essence." Without adopting the extravagancies of contemporaries who deny any human "essence" as an innate constraint upon development, we have to take more seriously than the ancients did man's extraordinary malleability and adaptability. It is no longer possible to share the romantic and optimistic beliefs of earlier humanisms in the innate goodness and benevolence of uncorrupted "human nature." That increasing knowledge of and sensitivity to humanistic masterpieces should *eo ipso* lead to an access of moral virtue seems too implausible ever to have been seriously maintained. And as much can be said of the simplistic conception of a unitary and sufficient power of Reason, illuminating indifferently all subjects of human concern.

Does this leave the temple of "Humanity" in ruins and as irrelevant to present concerns as the abandoned Acropolis? We may be confident at least that a revived humanism, relevant to our own pressing problems, rather than to those of

Athenian gentlemen or Renaissance courtiers, will need to be significantly different from its predecessors.

In what follows, I shall offer some tentative suggestions for the construction of such a humanism. I shall take as a starting point some summary reflections upon the limitations of "scientific method." The power and efficacy of science and its step-sister technology need no further acknowledgment; but if we are to avoid the polar heresies of scientolatry and scientophobia, we need to remind ourselves of what science and technology cannot accomplish. In this way, the opportunities and tasks of a regenerated humanistic education may become somewhat plainer.

4. *Scientific investigation is conducted within special and characteristic "perspectives" whose influence it seeks progressively to neutralize.* The great-great-grandson of Macaulay's schoolboy can be counted upon to "know" that scientific method consists in a resolute suppression of bias and prejudice, for the sake of a resolute confrontation with "neutral facts." They are the raw material for generalizations and higher-level theories, themselves ultimately judged and evaluated by their power to predict novel "facts." An adherent to this neo-empiricist myth, for that is what it is, is unlikely to be unaware of the ways in which the "data" and "principles" of science are necessarily mediated by a man-made apparatus of record and conceptualization. The fruits of scientific investigation are more like a map than like a photograph — and even the most "realistic" photograph is intelligible only in terms of a system of representation.

I propose to use the expression 'perspective' in the following technical senses. By a 'perspective' I mean a systematic repertoire of devices for recording and formulating data and conclusions relating to them. (The "Mercator Projection" is a paradigm instance of a 'perspective' in the intended sense.)

A perspective consists at least of the following: (i) observers and instruments of "observation" or record, together with rules for their use, (ii) a language or other symbolism of representation, (iii) associated, but not necessarily formulated practices (customs, procedures, routines) by means of which the aforementioned observers, instruments and the language used by them are "applied" or "activated," (iv) a stock of "given," unquestioned deliverances couched in the language of the perspective in question — i.e. the dogmas, axioms, common-places, etc. of the system in question.

We might think of a "perspective" as a relatively organized way of grasping (recording, perhaps also interpreting, explaining) some aspects of "experience" — a complex strategic apparatus compounded of material instruments and means of representation, together with relevant beliefs and modes of action, for coming to terms with part of the "world." Maps, chronologies, geometries, scientific theories, myths, *Weltanschauungen*, eschatologies, theodicies, can all be regarded as aspects or features of "perspectives" in the intended sense.

Perspectives are inevitably, to some degree, arbitrary: we can usually distinguish within them what is settled in advance of a particular report or record, what is "a priori" for the user of *that* system. Since the features of a report ascribable to the perspective itself are easily confused with features emanating from beyond it (the "information," "insight," etc. we seek) all perspectives are intrinsically error-prone.

Science seeks to mitigate the intrinsic fallibility of each cognitive perspective (a given coordinate system, a given mode of symbolization, a given methodology) by seeking invariants of wide classes of perspectives. By seeking what remains true of each perspective of the preferred class, it overcomes the arbitrariness that distinguishes one perspective from another. In this way, there emerges an etherealized limiting conception of the goal of "scientific objectivity" — a view of the universe "as it actually is," no longer transformed or distorted by an arbitrary "perspective," the universe seen from no particular standpoint, *sub specie aeternitatis*, in a God's eye view.

The conception of "scientific objectivity" I have sketched has an appealing sublimity: on its own ground it has extraordinary achievements to its credit. But the costs of the splendid success of "the Scientific Outlook" are substantial, among them the principled and systematic exclusion of much that might be said to constitute *human* perspectives — the reference-systems in terms of which men and women live and cope with their problems as historically situated and conditioned persons.

5. *The "perspective" in terms of which a human being makes such sense as he can, at a given time, of his "world" evades the jurisdiction of any scientific "framework."* A human being, located at a particular point in space-time, has repeated occasion to make reports that are excluded from an austerely scientific perspective. Such remarks, as "I am hungry," "I remember his dying words," "That looks horrible," "Help!," "I am dying of boredom," are, in the first place, essentially *self-referential*, with the first-person pronoun used or understood. More generally, the language of ordinary life ("ordinary language" for short) would be paralyzed in the absence of such "token-reflexive" words as 'I,' 'You,' 'here,' 'now,' 'later.' Our need for such context-dependent words corresponds to the inhomogeneity and anisotropy of the changing "world" in which each of us lives. Unlike the abstract and idealized "world" of physics, in which all spaces and times are intrinsically indistinguishable, "*my* world" has a center, a unique but shifting "origin." This is true, of course, even of a scientific perspective, since science is after·all an activity of space- and time-bound persons. But science, as I have said, drives constantly toward the neutralization of the "personal equation," everything that marks a "report" as emanating from this or that particular human source. In ordinary life, however, the personal reference is of the essence: it is not a matter of indifference to a man that *he* must die, nor is the poignancy of this thought captured by the "impersonal" translation "John Smith will die." Closely related to these points is the reality, in the personal world, of change and hence the indispensability in ordinary language of tenses, which the essentially timeless cosmos of science does not capture. The '*t*' of physics, whether in its contemporary or its older form, is simply one more variable in a four-dimensional space.

An equally radical point of contrast between "ordinary language" and the "language of science" is the use of the former to express in multifarious ways, feelings, attitudes, intentions, fears, aspirations, the whole gamut of the so-called "non-cognitive" uses of language. Here, as in the case of the supposedly "objective" transformation of utterances with essentially personal and "indexical" reference, the step from an expression of some human parameter (attitude, intention, feeling or whatever) to an assertion about that parameter involves an attenuation of context and deflection of function.

Try to imagine a "society" whose members, prevented by the poverty of their language from expressing "human parameters," are limited to assertions *about* intentions and the like, and whose only "value" is truth. It is, of course, inconceivable that human beings could be mere reporters and communicators, while never enacting the passions they purported to be observing: in the absence of expressive symbolism, there would be nothing to report.

The "personal language" of a space-and-time-bound person, that part of the common language of the group(s) to which he belongs that is always formulated, whether explicitly or implicitly, in the first person singular or plural, is incommensurable with the language of science. Scientific talk sometimes applies to persons, for a human body is still a physical body, but to vast numbers of matters of personal concern scientific pronouncements are simply irrelevant.

Yet the "world" of an actual individual, however variable and incoherent, is not a chaos: it, too, may properly be brought under the rubric of "perspective," so that we might speak of "P(personal)-perspectives." For here, too, we can discern systematic and relatively stable repertoires of expression (a language, together with paralinguistic "gestures"), resting upon temporarily accepted "commonplaces" and activated by routines of action. It is of decisive importance for the role of humanistic education that such a "world" or "perspective" is unequally illuminated by consciousness and self-awareness: we hardly needed Freud to remind us that much of what we do is masked.

P-perspectives intersect G(group)-perspectives. For the language that is a central component of a P-perspective, no matter how self-centered, idiosyncratic or solipsistic, is necessarily a social instrument: every utterance is potentially intelligible to another person; every expression might "make sense" to a companion; as integral as the reference to "I" are the correlative references to "We," "You" and "They." From a linguistic standpoint: the first person singular depends for its use and function upon a contrast (opposition) with the other pronouns. P-perspectives and the G-perspectives that overlap them might be called H(human)-perspectives.

6. *A distinctive task of humanistic criticism is the delineation and articulation of "human perspectives."* I use "criticism" here in a broad sense to cover all the second-order activities of a humanistic scholar, as distinct from those of a humanistic creator. Poets, speculative philosophers, and historians, are inter alia, engaged in *presenting* possible worlds, "perspectives," or aspects of them, but not

in the relatively detached and "distanced" mode of criticism. Roughly speaking, a creator shows by and in his constructions what H-perspectives are and might be; the critical commentator tries to make plainer, within a metalanguage, what the artist is doing in his object-language. The poet presents us with an expressive representation of the human condition; the critic sometimes succeeds in helping us to understand better, to see more clearly, what the poet is doing.

I am using "perspective" here, as before, to stand for a relatively organized constellation of axioms, assumptions, beliefs, expectations, aspirations, attachments, obligations, preferences, evaluations, characteristically crystallized in a distinctive sub-language, conceptual structures and set of practices. That is to say, a way of "seeing the world" and acting within it as thus perceived. All such perspectives, I have claimed above, are, to some degree, arbitrary, as embodying some one of many conceivable alternative choices of "modes of representation" that constitute what is "a priori" for the committed *user* of the framework in question. Thus any "framework" can in principle be judged as more or less "effective," "adequate" or "valid" by comparison with other frameworks, however difficult this task of comparative evaluation may be for somebody whose very language and modes of perception and thought are determined by a framework that is his own.

The special perspectives that I have distinguished as "human" and particularly those that I have called "personal" (the world-views of individuals) are particularly amenable to criticism, even when possessed by the most dogmatic and authoritarian subjects, for the following reasons. The "ordinary language" that is the central component of a P-framework has the important resources of *reflexiveness* (the uses of language to refer to that language itself) and *suspension of assertion* (the expression of possibilities that the language-user can consider prior to commitment or choice). The first of these is crucial for the possibility of the modes of self-criticism manifested in shame, remorse, regret and, less dramatically, in desire for change and improvement; the second is crucial for the possibility of understanding the standpoint of those other symbol-users with whom the user of a particular P-framework necessarily interacts, through his use of a shared language. Each "idiolect" is a variation upon the common dialect of a group.

Another name for what I have rather pompously called "suspension of assertion" is imagination, whether in its humblest or most spectacular exercises. In order to choose anything at all, we have to draw upon our linguistic framework's capacity to represent the possible, delineate in advance of the choice, "what it would be like if we were to choose otherwise." The same is true, more poignantly, of the constant demands of others that we shall "try to understand them." We succeed in this, however imperfectly, to the degree that we make the other's world "our own" by projecting part of his life-space into ours: we understand others' expressions of intention, attitude, feeling, belief by knowing what we might express by those words in their situation; conversely we achieve higher self-awareness of our own intentions, etc. by learning, in speech as well as in non-verbal interaction, how our "public" expressions are interpreted. Such imaginative participation depends strongly upon that temporary "bracketing" or "suspension of disbelief" whose highest elaboration in literature is continuous with its humbler exercises in the market place.

Given that every "private world" (P-perspective) is unequally illuminated and that the drive toward illumination of its darker regions ("self-awareness") is as real as the contrary drive to suppress and to conceal by "masking," we can appreciate the centrality for self-education of exercises in imaginative participation. Since every private world, however solipsistic and egoistic, is intersected and sustained by a social world, imaginative participation via suspension of disbelief (and hence also suspension of assertion and action) is a potent means for insight into our own worlds. By entering empathically into the expressive utterances of others , and thereby making them our own as possibilities, we deepen our insight into our individual life-space. Through the words and deeds of others we apprehend the grain of our own life-space: we see more clearly what we can and might do. We learn not so much what we *must* do in order to be saved, which implies commitment to an accepted framework, as what we *might* do to be "saved" if we were the kind of person to be "saved" in *that* way. Imaginative participation broadens the range of available, because perceived, choices: that is why all creative literature and art is necessarily innovative, no matter how closely it might seem to adhere to traditional forms.

7. *A basic method for progress toward "perspectival articulation" is "imaginative participation" in works of art.* I have tentatively defined a personal perspective as the system of concepts, axiomatic assumptions and beliefs, modes of expression of feelings and attitudes, strategies for problem solving, etc., which are, at a certain time, "given" for a particular person: the cognitive, perceptual, expressive and volitional structure in terms of which, and by reference to which, he "sizes up" *his* "world." Suppose now that we are interested in getting detailed knowledge of such a "perspective" — or, what comes to the same thing, in understanding the other's point of view, his "attitude towards life," how should we best proceed? I assume that, unless our interest is specifically biographical, we shall focus upon the features of the personal perspective that are shareable, as constituting part of a *group*-perspective.

It is natural, especially for scholars who are professional writers and talkers, to think immediately of verbalizing the perspectival apparatus to which I alluded at the start of the last paragraph. Indispensable as such a verbal formulation of a perspective may be, it has the crucial weakness, all too familiar to any teacher of a humanistic subject, of stereotyping and sloganizing its object. Consider the examinee's task of, say, characterizing Tolstoy's perspective as expressed in *War and Peace:* the best we can hope from even the most sensitive and literate student will be a set of rapid generalities, all-too-true alas, about Tolstoy's commitment to historical determinism, his anxieties about sexual attraction, and so on. The task is, in principle, too difficult to be solved in that way.

"Installed Figures", 5' 1" x 26" (Woodcu

The radical untranslatability that we here encounter as an obstacle to the interpreter's charge is present in all art, but also throughout ordinary life: that Pierre's love for Natasha is not the same as Andrei's even though they fall "under the same description" would be a problem for Natasha, too, if she really existed. The difficulty is partly in the higher "multiplicity" of the action *vis-a-vis* its verbal description: the near impossibility of capturing in words the difference between the way a child strokes a cat and its mother does. But it is not only that. So far as it goes, the expression "a jealous retort" may be perfectly correct in its application to a particular act and yet, as we say, mean nothing to somebody who has never been jealous, for whom jealousy is only a name. (There is such a thing as feeling blindness as well as colour blindness.) To the extent that actions are the enactments of attitudes, we can fully understand them only by some process of imaginative participation (or empathic communion) in which, by rehearsing the attitude, we come to know it from the inside.

We must avoid the blunder of supposing that because the rhythmic and qualitative aspects of action are not to be captured in verbal description, they are, therefore, inexpressible and somehow ineffable. On the contrary, they are typically expressed, and the attitudes, motives, intentions, etc. that inform them are essentially expressible. Enough mystifying rubbish has been written about "Einfuhlung" or "empathy" to make us wary. Nevertheless, I submit that it would be gratuitous reductionist folly to deny it a central place in the enterprise of humanistic criticism.

Suppose, then, that we have the power of imaginative participation or empathic understanding into the actions and works of others; what should we call its outcome? I think knowledge is a plausible answer — the kind of knowledge that we have when we see how a painting looks, or understand, perhaps after reading some fictional depiction of jealous actions, what jealousy is like. Those who would deny this the name of knowledge are, without knowing it, in the grip of an ancient prejudice that insists on restricting knowledge to what can be talked about.

8. *Further Problems.* My positive suggestions might be taken as a preliminary sketch for a philosophy of the Humanities, in whose absence discussion concerning the educational functions of those studies and their bearings upon the pressing social and personal problems created by science and technology is likely to remain confused, incoherent and ineffective.

Even as a sketch, it has serious deficiencies. I have, for instance, stressed the centrality of literary texts for a revived program of humanistic education, at the expense of historical and philosophical studies, which deserve at least equal prominence. Considered even in its application to literature, my emphasis upon the notion of a "human perspective" runs the risk of treating every work of art as a didactic allegory, which is far from my intention. Indeed, emphasis upon a "perspective" might well be regarded as one-sided, as tending to elide the importance of human projects and predicaments. A professional philosopher may well shy away from my suggestion, toward the end, that knowledge need not be mediated by discursive symbolism. Teachers might well wonder about the bearings of my analysis upon curricula and courses of study. And so on. Some of these problems I hope to discuss eventually in an ampler context.

(Also published in W. R. Niblett, ed., *The Future Contribution of the Humanities to Higher Education,* University of London Press, 1973).

Military Technology: A Problem of Control

John Kenneth Galbraith
Harvard University

The importance of military spending in the economy—half the federal budget, about one-tenth of the total economic product, I need not stress. Though much attention is focused upon it, this bloodless economic side is not, I venture to think, the important feature. The important feature is the peculiar constitutional and bureaucratic arrangements which govern this economic activity.

In our ordinary economic arrangements we think of the individual as instructing the market by his purchases, the market, in turn, instructing the producing firm. Thus economic life is controlled. This the textbooks celebrate. And where public expenditures are concerned, the young are still taught that the legislature reflects the will of the citizen to the Executive. The Executive, in turn, effects that will.

I have argued that with industrial development—with advanced technology, high organization, large and rigid commitments of capital—power *tends* to pass to the producing organization—to the modern large corporation. Not the consumer but General Motors tends to be the source of the original decision on the modern automobile. If the consumer is reluctant he is persuaded—to a point at least.

This part of my case has not escaped argument. Dissent raises its head everywhere these days. But where military goods are concerned one encounters little or no argument. Here, it is agreed, the historic economic and constitutional sequence *is* reversed. The citizen does not instruct the legislature and the legislature the Pentagon and its associated industries. No one wants to be that naive. Vanity becomes the ally of truth. It is agreed that the services and the weapons manufacturers decide what they want or need. They then instruct the Congress. The Congress, led by the military bureaucrats and sycophants among its members, hastens to comply. The citizen plays no role except to pay the bill. As I say, these matters are not subject to serious dispute, those with a special capacity to believe in fairy tales apart.

The power that has brought this remarkable reversal—has assumed this authority—has, of course, been well identified. It is the military services acting individually or in association through the Department of Defense and the large military contractors. The latter, an important point, are few in number and highly specialized in the service to the military. In 1968, a hundred large firms had more than two-thirds (67.4 percent) of all defense business. Of these, General Dynamics and Lockheed had more than the smallest fifty. A dozen firms specializing more or less completely on military business—McDonnell Douglas, General Dynamics, Lockheed, United Aircraft—together with General Electric and A.T.&T. had a third of all business. For most business firms business is inconsequential except as it affects prices, labor and material supply—and taxes. The common belief that all business benefits from weapons orders is quite wrong. For a few it is a rewarding source of business. The great multitude of business firms pay. The regional concentration, I might add, is equally high; in 1967 a third of all contracts went to California, New York, and Texas. Ten states received two-thirds. And no one should be misled by the argument that this picture is substantially altered by the distribution of subcontracts.

One must not think of the military power—the association of the military and the defense firms—in conspiratorial terms. It reflects an intimate but largely open association based on a solid community of bureaucratic and pecuniary interest. The services seek the weapons; the suppliers find it profitable to supply them. The factors which accord plenary power of decision to the military and the defense plants, and which exclude effective interference by the Congress and the public, are quite commonplace. Nothing devious or wicked is involved. The following are the factors which sustain the military power.

First: There is the use of fear. This, of course, is most important. Anything which relates to war, and equally to nuclear weapons and nuclear conflict, touches a deeply sensitive public nerve. This is easily played on. The technique is to say, in effect, "Give us what we ask, do as we propose, or you will be in mortal danger of nuclear annihilation." In this respect one must pause to pay tribute to Secretary of Defense Laird. He has shown himself, on this matter, to have a very high learning skill.

Second: There is the monopoly, or near monopoly, of technical and intelligence information by the services, their suppliers, and the intelligence community. This monopoly, in turn, is protected by classification. This allows the military power to exclude the lay critic, including the legislator, as uninformed. But even the best scientist can be excluded on the grounds that he is not fully informed on the latest secret technology—or does not have the latest knowledge on what the Soviets or the Chinese are up to. Here too the new administration has been very apt. If Secretary Laird deserves a special word of commendation on the way he has learned to use fear, Under Secretary Packard must be congratulated on the speed with which he has learned to discount criticism as inadequately informed of the latest secrets.

Third: There is the role of the single-firm supplier and the negotiated contract. These are largely inevitable with high technology. One cannot let out the MIRV to competitive bidding in the manner of mules and muskets. In fiscal 1968, as the Joint Economic Committee has revealed, 60 per cent of defense contracts were with firms that were the sole source of supply. Most of the remainder were awarded by negotiated bidding. Competitive bidding—11.5 percent of the total—was nearly negligible. With single-firm supply, and in lesser degree with negotiated supply, opposition of interest between buyer and seller disappears. The buyer is as interested in the survival and well-being of the seller as is the seller himself. No one will enter this Elysium to cut prices, offer better work, earlier deliveries or cry favoritism. That is because there is no other seller. The situation, if I may be permitted the word, is cozy.

Fourth: There is the fiction that the specialized arms contractor is separate from the services. The one is in the public sector. The other is private enterprise. As Professor Murray Weidenbaum (the notable authority on these matters), as well as others, have pointed out, the dividing line between the Services and their specialized suppliers exists mostly in the imagination. Where a corporation does all (or nearly all) of its business with the Department of Defense; uses much plant owned by the government; gets its working capital in the form of progress payments from the government; does not need to worry about competitors for it is the sole source of supply; accepts extensive guidance from the Pentagon on its management; is subject to detailed rules as to its accounting; and is extensively staffed by former service personnel, only the remarkable flexibility of the English language allows us to call it a private enterprise. Yet this is not an exceptional case, but a common one. General Dynamics, Lockheed, North American-Rockwell and such are public extensions of the bureaucracy. Yet the myth that they are private allows a good deal of freedom in pressing the case for weapons, encouraging unions and politicians to do so, supporting organizations as the Air Force Association which do so, allowing executives to do so, and otherwise protecting the military power. We have an amiable arrangement by which the defense firms, though part of the public bureaucracy, are largely exempt from its political and other constraints.

Fifth: This is a more subtle point. For a long period during the fifties and sixties during which the military power was consolidating its position, military expenditures had a highly functional role in the economy. They sustained employment; they also supported, as no other expenditures do, a high technical dynamic. And there was no wholly satisfactory substitute. More specifically, a high federal budget, supported by the corporate and progressive personal income tax, both of which increased more than proportionally with increasing income and reduced themselves more than proportionally if income faltered, built a high element of stability into the system. And the scientific and technical character of this outlay encouraged the expansion of the educational and research plant and employed its graduates. It was long a commonplace of Keynesian economics that civilian spending, similarly supported by a progressive tax system, would serve just as well as military spending. This argument which, alas, I have used myself on occasion was, I am now persuaded, wrong—an exercise in apologetics. Civilian spending does not evoke the same support on a large scale. (Even in these enlightened days I am told that Representative Rivers prefers naval ships to the Job Corps.) And although it is now hard to remember, the civilian pressures on the federal budget until recent times were not extreme. Taxes were reduced in 1964 because the pressures to spend were not sufficient to offset tax collections at a high level of output—to neutralize the so-called fiscal drag. And civilian welfare spending does not support the same range of scientific and technical activities, or the related institutions, as does military spending. On a wide range of matters—electronics, air transport, computer systems, atomic energy—military appropriations paid for development costs too great or too risky to be undertaken by private firms. They served as a kind of honorary non-socialism.

Sixth and finally: There is the capacity—a notable phenomenon of our time—for organization, bureaucracy, to create its own truth—the truth that serves its purpose. The most remarkable times, of course, has been Vietnam. The achievements of bureaucratic truth here have been breathtaking. An essentially civilian conflict between the Vietnamese has been converted into an international conflict with a rich ideological portent for all mankind. South Vietnamese dictators of flagrantly repressive instincts have been converted into incipient Jeffersonians holding aloft the banners of an Asian democracy. Wholesale larceny in Saigon has become an indispensable aspect of free institutions. One of the world's most desultory and impermanent armies—with desertion rates running around 100,000 a year—was made, always potentially, into a paragon of martial vigor. Airplanes episodically bombing open acreage or dense jungle became an impenetrable barrier to men walking along the ground. An infinity of reverses, losses, and defeats became victories deeply in disguise. There was nothing, or not much, that was cynical in this effort. For, for those who accept bureaucratic truth, it is the unbelievers who look confused, perverse, and very wrong. Throughout the course of the war there was bitter anger in Saigon and here in Washington over the inability of numerous people—journalists, professors, and others—to see military operations, the Saigon government, the pacification program, the South Vietnam army in the same rosy light as did the bureaucracy. Why couldn't all sensible people be the indignant instruments of the official belief—like Joe Alsop? (If I may pay tribute to the Edward Gibbon of the Vietcong.)

An equally spectacular set of bureaucratic truths has been created to serve the military power—and its weapons procurement. There is the military doctrine that whatever the dangers of a continued weapons race with the Soviet Union, these are less than any agreement that offers any perceptible opening for violation. Since no agreement can be watertight this largely protects the weapons industry from any effort at control. There is the belief that the conflict with communism is man's ultimate battle. Accordingly, no one would hesitate to destroy all life if communism seems seriously a threat. This belief allows acceptance of the arms race and the production of the requisite weapons no matter how dangerous. The present ideological differences between industrial systems will almost certainly look very different and possibly rather trivial from a perspective of fifty or a hundred years hence if we survive. Such thoughts are eccentric. There is also the belief that national interest is total, that of man inconsequential. So even the prospect of total death and destruction does not deter us from developing new weapons systems if some threat of national interest can be identified in the outcome. We can accept 75 million casualties if it forces the opposition to accept 150 million. We can agree with Senator Richard Russell that, if only one man and one woman are to be left on earth, they should be Americans. (Not from any particular part of the country, just Americans.) We can make it part of the case for the Manned Orbiting Laboratory (MOL) that it would maintain the American position up in space in the event of total devastation from Maine to California. Such is the power of bureaucratic truth that these things are widely accepted. And being accepted they sustain the military power.

What now should be our response? How do we get the power under control?

Our response must be in relation to the sources of power. Again for purposes of compressing this discussion, let me list specific points:

1. Everyone must know that fear is deployed as a weapon. So we must resist it. I am not a supporter of unilateral disarmament. I assume that the Soviets also have their military power sustained by its bureaucratic beliefs. But we must look at the problem calmly. We must never again be stampeded into blind voting for military budgets. These, as a practical matter, are as likely to serve the bureaucratic goals of the military power and the pecuniary goals of the contractors as they do the balance of terror with the Soviets. And we must ascertain which.

2. That part of the military budget that serves the balance of terror can be reduced only with negotiations with the Soviets. As Charles Schultze and others have pointed out, however, this is a relatively small part of the military budget. The rest serves the goals of the military power and the interests of the suppliers. This can be curtailed. But it can only be curtailed if there is a vigorous reassertion of congressional power. Obviously this will not happen if sycophants of the military remain the final word on military appropriations. The Congress has the choice of serving the people in accordance with constitutional design or serving Senator Russell and Representative Rivers in accordance with past habit.

3. Informed technical and scientific judgment must be brought to bear on the foregoing questions. This means that the Congress must equip itself with the very best of independent scientific judgment. And the men so mobilized must not be denied access to scientific and intelligence information. I believe that on military matters there should be a panel of scientists, a Military Audit Commission, responsible only to the Congress—and not necessarily including Edward Teller—to be a source of continuing and informed advice on military needs—and equally on military non-needs.

4. We must, as grown-up people, abandon now the myth that the big defense contractors are something separate from the public bureaucracy. They must be recognized for what they are—a part of the public establishment. Perhaps one day soon a further step should be taken. Perhaps any firm which, over a five-year period, has done more than 75 percent of its business with the Defense Department, should be made a full public corporation with all stock in public hands. No one will make the case that this is an assault on private enterprise. These firms are private only in the imagination. The action would ensure that such firms are held to strict standards of public responsibility in their political and other activities and expenditures. It would exclude the kind of conspiracy uncovered in the Lockheed case. It would help prevent private enrichment at public expense. In light of the recent performance of the big defense contractors, no one would wish to argue that it would detract from efficiency. And the 75 percent rule would encourage firms that wish to avoid nationalization to diversify into civilian production. Needless to say, the 75 percent rule should be applicable to the defense units of the conglomerates. Perhaps to press this reform now would direct energies from more needed tasks. Let us, however, put it on the agenda.

5. Finally, it must be recognized that the big defense budgets of the fifties were a unique response to the conditions of that time. Then there were the deep fears generated by the cold war, the seeming unity of the communist world, and, at least in comparison with present circumstances, the seeming lack of urgency of domestic requirements. All this has now changed. We have a wide range of tacit understandings with the Soviets; we have come to understand that the average Soviet citizen—in this respect like the average American voter—is unresponsive to the idea of nuclear annihilation. The communist world has split into quarrelling factions. I am enchanted to reflect on the Soviet staff studies of the military potential of the Czech army in case of war. Perhaps, as I have said elsewhere, we have here the explanation of their odd passion for the Egyptians. And as all philosophers of the commonplace concede, we have the terrible urgency of civilian needs—of the cities, the environment, transportation, education, housing, indeed wherever we look. It is now even agreed to as where the danger to American democracy lies. It is from the starvation of our public services, particularly in our big cities, here at home.

Let me make one final point. Our concern here is not with inefficiency in military procurement. Nor is it with graft. These divert attention from the main point. And this is not a crusade against military men—against our fellow citizens in uniform. Soldiers were never meant to be commercial accessories of General Dynamics. It would horrify the great captains of American arms of past generations to discover that their successors are by way of becoming commercial accessories of Lockheed Aircraft Corporation.

The matter for concern is with the military power—a power that has passed from the public and the Congress to the Pentagon and its suppliers. And our concern is with the consequences—with the bloated budgets and bizarre bureau-

100

cratic truths that result. The point is important for it suggests that the restoration of power to the Congress is not a sectarian political hook. It is one for all who respect traditional political and constitutional processes.

(Also published in Charles R. Beitz and Theodore Herman, eds., *Peace and War*, San Francisco, Calif., W. H. Freeman and Co., 1973.)

Technology and New Institutions in Human Settlements

John P. Eberhard
American Institute of Architects Research Corporation
Washington, D.C.

The rate by which the human density of our planet is being increased is alarming. In the span of one week, a million more children are born; in a year, fifty million; and by the end of the century our population will have been exploded by 1.35 billion. Faced with this kind of density we will have to provide an additional one billion housing units in the next twenty-seven years; two hundred million to replace those units that will wear out or be destroyed during these three decades and two hundred million which we should build to replace the slum dwellings presently included in the world inventory of five hundred million.

In addition to having to face the problems related to density, all of us are equally overwhelmed by the intensity of the pace of life, by the rate of publication and by the rate of production of new inventions. This is all too apparent to me when I realize that since last year 38,000 books have been published and I still haven't read the three books I had marked high priority for the Christmas holidays. Since 1963 when I left my position as a visiting faculty member of M.I.T.'s Sloan School of Management to work in Washington, the amount of knowledge in the world has doubled, as it has in every decade of this century. At that time I thought I was reasonably well-educated even though I had probably absorbed less than one percent of the available knowledge. In the past ten years although I have learned more, I am, relatively speaking, fifty percent less well-educated to deal with the increasing complexity of knowledge.

To meet the demands of the density, intensity and complexity of life, we should be innovating at a prolific rate, if we are just to mark time. Yet, as Don Schon points out, most institutions are expending the majority of their energies in attempting to preserve their stable state. Peter Drucker believes that we are entering a period of "discontinuity" — a time during which we cannot predict the immediate future by extrapolating from the technical developments of the recent past. Daniel Bell and others say we are entering a period which we should call the "post-industrial" period, when the institutions and rule systems set up during the so-called period of "industrial revolution" will no longer be valid.

Most of us feel these changes in our bones and we react in different ways. Some of us try desperately to hang on to concepts, institutions, and approaches we have inherited. Some of us try to shake off all of the traditions and intellectual baggage of the past as we set out to build a new kind of society. But most of us are caught somewhere in between—not ready to forsake the past, unsure that what is new is really better, and hoping that somewhere, somehow, order and stability will enter our lives, so that it will be easier to cope with the problems of density, the pressures of intensity, and the difficulties of complexity.

If we are to achieve such order and stability, we need to recognize the nature of the global issues of urbanization which we face. The urban context is growing, despite the fact that the central city is declining. We are now more conscious than ever of the causes of this decline, of the negative impact that the growth of the suburbs has had on the quality of life and on the physical environment of our cities; we are concerned about the rot and decay, and about the climate of fear that pervades in our cities; we are sensitive to the overwhelming economic problems that the cities are experiencing because of the exodus of the white and middle-class populations to the suburbs. We are also aware that there is sufficient vacant land in our major cities to provide for all of the new housing stock we will need in these next three decades, if it were seen as a resource to be properly managed in the public interest, and that there is enough land to recreate the kind and character of urban existence that would enhance the lives of rich and poor alike.

Indications are that there is a renewed interest in finding solutions to the problems of the cities and in recreating their vitality. It is here that I see the challenge and the next big opportunity for the scientific, technological and professional communities. Having recently returned to Washington after an absence of five years, it is evident to me that regardless of how "Watergate" is resolved, there is a shift in attitude towards certain programs and there are new priorities for the allocation of funds. The space program has little or no chance of being any more than a modest effort in the immediate future; weapons systems still have a lot of champions in the halls of power, but not in the voting booths. Atomic energy is being sharply criticized for its lack of progress in meeting the nation's energy needs, while being busy producing an overkill capacity for the military machine.

During the past two decades we invested about ninety percent of our Federal R&D dollars in space programs, weapons systems, and atomic energy. In the process we created a giant military-space-industrial complex, a large and overly proud house of science, and an era in higher education that made the research dollar the arbiter of intellectual development. No where is this linked set of developments more in evidence than in the Massachusetts Institute of Technology. With students drawn from the top two percent of their high school classes as candidates for increasingly more complex degrees, the intensity of their educational patterns was harnessed to big science and high technology in the service of those three areas. Now M.I.T. and other universities, along with the rest of the science and technology community, are having financial problems and are searching for new targets of opportunity.

"Car Door #3", 10" x 48" x 40" closed, 10" x 48" x 80" open (Acry

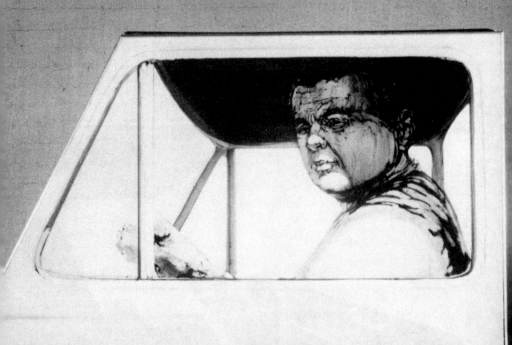

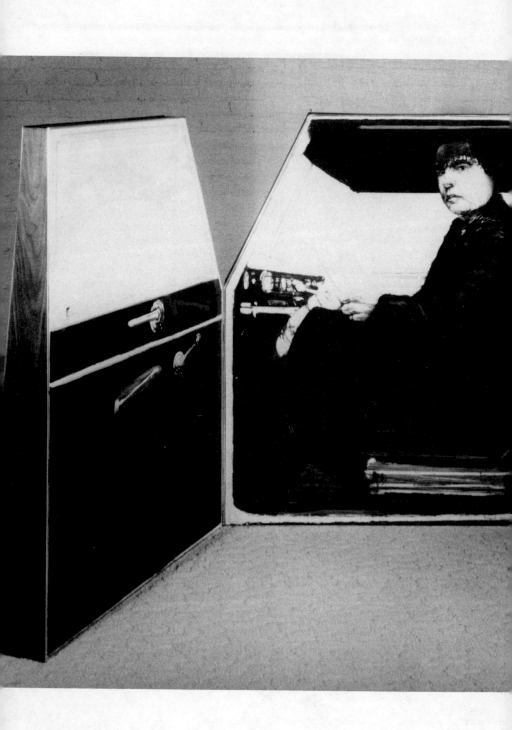

Ecological and environmental pollution will probably be one of these targets, as will the energy crisis and the impending shortage of critical materials, especially the exotic metals. But the underlying target of all three of these is the urban environment. It is there that we are fouling our nest the most, in terms of pollution. It is there that the energy crisis will have the greatest impact. It is there that the greatest potential for innovation in new high-technology systems will be found.

Anyone who is associated with one of the traditional industries that has been supplying materials and components to the construction industry, or who is a part of one of the traditional professional groups that has been supplying the design and engineering talents required to produce urban places, had better prepare for an invasion. Not for a few minor innovations like plastic pipe or turnkey contracts, but for a real upheaval—a new ball game with new players and new roles.

A new ball game with new players will mean a period of major and minor innovations, including both process and product. There will be new skills to learn, new institutions to create, new kinds of resource allocations and new regulations and procedures. It will also mean that new questions will have to be asked, such as: "How can educational institutions hope to organize programs, courses, and teaching methods to prepare the next generation to deal with such a period of innovation?" "How can elementary and secondary education hope to pave the way for young minds so that they will be ready to absorb new concepts at a faster rate?" "How can all of us who still need to be re-educated, find the time and the opportunity for this education, so that we can adjust both personally and professionally?"

It won't be easy, but it could be an exciting time to be around. It could be a time of renewed involvement of the young in a positive cause with which they felt some sympathy. It could breathe new life into some tired institutions. It could offer new hope and encouragement to those now trapped in the cycle of poverty and unemployment in our urban centers. It could be a time of new ventures right here on Earth of the kind we really haven't seen yet in this century.

But first, let's look back at what happened in the closing decades of the last century, as a model of urban change. After the Civil War, or War between the States, was over and the period of reconstruction began, we had a burst of inventions and innovations which were primarily addressed to the urban condition. The emergence of the industrial revolution, the rise of the corporate form of business as a way of gaining investment capital and the flowering of science and the mechanical arts produced a confluence of forces that shaped our cities into new forms never previously imagined.

For centuries we had built our cities across the world from simple materials, either exactly as they were found in nature, like wood and stone, or from materials in only slightly revised form, like brick and concrete. We needed only limited forms of energy to make those older cities work—coal and wood for fireplaces, candles or whale oil for lights, and animals to carry us or pull us around a little faster than our feet could carry us. We made our houses by hand with care in order that we could pass them on to our sons and grandsons. We built great cathedrals from prefabricated parts that took decades to fashion in stone, brick, and stained glass. We used water power and wind power to operate crude machines in small shops. Mostly, we used simple hand tools in cottages where we

106

lived. Life was slow, work was hard, and pleasures were more spontaneous than organized.

By 1875, the railroad had begun to change the pace of progress. The steam engine found its way into ships and into the factory as well as the railroad. Work began to be organized on a larger scale so that people flocked to the towns to man the factories. We discovered how to extract fuel from oil that could be burned in another invention called the internal combustion engine, and in time we invented contraptions that could use the mechanical power of such engines to carry us around. Man begat the automobile, the truck, and the bus. He discovered a phenomenon called electricity. We learned how to make it and how to transmit it and then all kinds of ways to use it. Man begat the electric light, and the street car, and the electric engine to work his machines. A man named Bell discovered how to use electricity to transmit voice signals over wires and man begat the telephone. We found ways to build a fire inside a device called a furnace and to transmit heat throughout buildings large and small by water pipes or ducts of air. Man begat central heating. We put our engineering skills to work on large hydraulic projects that began to substitute the toilet for the privy in the back yard, the kitchen for the water pump or well, and the bathtub for Saturday night cleaning in the kitchen tub. And man begat the bathroom in its modern form. We put together some large corporations to make steel for the railroads and as the price of urban land went higher and higher we discovered how to make steel frames for buildings, so that they could go higher and higher. And man begat the skyscraper and the elevator to make the buildings useable.

This breathtaking list of innovations, which changed the city, were all reduced to practice—patented in their basic form—between 1880 and 1892. In just twelve years, the fabric of the modern city was created. Since that time, with all that has transpired, there has not been another invention that has changed the fabric of the city. Radio and television didn't alter urban design or the physical shape of cities. The airplane connected cities but didn't change them. What I have called the second generation of urban hardware is based on those eight inventions and their subsequent product improvements. Since 1892, these eight inventions have been built into the institutional framework of our cities. Look at the organization of our building codes or the organizational structure of local city government. You'll find sections on structural safety with a steel handbook, elevator safety, sewer and water, roads and automobiles, heating device inspectors, etc. The professional societies—groups like the AIA, ASTU, ASME, HPPA, the original Society of Electrical Engineers—were all formed a few years before or after these twelve years of invention. Even our universities are organized around engineering programs that represent the concepts of these eight inventions, and most of our engineering schools were founded about the same period in time as the twelve years of invention.

To me this history suggests that we will first have to provide institutional innovations that educate, nourish, and encourage new kinds of professionals, new problem statements, new market opportunities, before we will be likely to produce a new set of hardware innovations in our cities. The existing institutions aren't likely to invent new notions that will displace their present markets. The history of inventions would tell us that. Railroads aren't an outgrowth of the horse and buggy business. Automobiles didn't represent a new product line for railroads. Airplanes are not another product of the auto manufacturer. Nylon and rayon were an invasion by the chemical industry into the textile industry. Computers invaded the market of business equipment manufacturers and are the products of new companies in a new industry. Necessity still tends to be the mother of invention, but companies and organizations in an established market seldom recognize the "necessity" that falls outside of their general concept of their user public.

This leads us to the question of what are the "necessities" for the urban context that might induce such new inventions once institutional innovations had occurred? It's easier to say what they are not than what they are. There is not basic need for steel framed skyscrapers, or elevators, for automobiles or telephones, even for central heating and plumbing. These are all recent innovations that were a response to the basic needs of mobility, information and communications, the need to supply raw materials and the subsequent removal of waste, the need for shelter to provide a controlled environment for work and play, and living the private part of our lives. Even houses aren't a necessity. They are a form of shelter, and only one among many the world has seen already. Surely we haven't exhausted our inventive genius in terms of alternatives. But we should also be as clear as possible about these "necessities" for mobility, communications, and shelter being the functional necessities which derive their performance needs from such original activities as education, health-care, and recreation.

To return to the subject of education for innovation, it seems clear to me that if we believe that we are about to go through a major upheaval in the way we design and build our cities, if we are going to have a whole new set of urban systems inventions, if these are likely to be developed by institutions which are new to the urban scene, that we should not continue to organize our university education as a form of apprenticeship to the existing professions and institutions. If we want to be cautious about these forecasts of change (which would be our normal tendency) we would not immediately change all of our educational programs, but we should be changing some of them. We must educate a whole new breed of professional who can bridge the gap between the different communities. This need for the removal of traditional disciplinary boundaries is true of engineering programs, law, the social sciences and the physical sciences, as well as architecture. I see signs of movement in this direction in some of our more venturesome universities.

The same kind of arguments might be advanced about elementary and secondary education. Our high schools seem to be organized to prepare apprentices for existing programs in universities. Traditional subjects like English Composition, American History, Chemistry, Physics and Mechanical Drawing are found in practically every high school. I have the feeling that the brightest and best high school students I know would be better educated for the world they are moving into, if they had courses dealing with how the world might deal with problems of density and what were the historical reasons for this development; how we can adapt to the intensity of our pace of life, ways to relax, to experience life, and to be creative; and how we can deal with complexity: some of the intellectual tools that we have, like the systems concept, and some of the hardware we have available to help, like the computer.

High school students with this kind of education, instead of narrow disciplinary training, should be better prepared for genuine educational experiences as contrasted to further training at universities and should be better prepared to be good citizens of an urban society even if they do not go on to a university.

Finally, there are the educational experiences in which all of us can participate, regardless of our age. Experiences that keep the mind open to new ideas and new potentials. In this way I believe that each of us can best deal with the density, intensity, and complexity of our current urban lives—not by retreating to the past, not by running away to some new utopia, but by constantly being educated through our experiences. This kind of education has no boundaries. It happens potentially in many places—while we are working, when we are visiting a new place, while we are talking with our children. Such an education is a state of mind, not a body of knowledge, not what is learned in an institutional setting. It is the kind of attitude that makes us open to change—and it is change that is our steady state.

The Craftsman: Some Reflections on Work in America

David E. Whisnant
Durham, North Carolina

"If I were a carpenter
And you were a lady
Would you marry me anyway,
Would you have my baby? . . .
If I worked my hands in wood
Would you still love me? . . ."

Tim Hardin, "If I Were a Carpenter"

"I thought you would despise me
If you knew what I'd really wanted to be."

Sir Claude Mulhammer to his wife in
T. S. Eliot's *The Confidential Clerk*

How does it feel to be a craftsman in industrial America?[1] What support and encouragement does one find, and against what odds does one strive? Are craftsmen — those quaint woodcarvers, leather and metal workers, potters and weavers one finds here and there — harbingers of a new post-industrial order, or at least of a healthy occupational pluralism hitherto unknown in our increasingly monolithic industrial society? Or are they anachronisms, dull-witted Mesozoic beats indulging themselves in nostalgia while waiting to be plasticized or paved over like the rest of America?

Assessing the craftsman's objective situation in American culture is complicated by ambiguities and ambivalences in the perspectives from which he is viewed. The craftsman, I have discovered through nearly ten years of craft work and talking with many active craftsmen, is usually viewed with a mixture of envy and condescension. Envy because doing beautiful work in one's own way and at one's own speed is possible for so few in our society. Condescension because one must be a bit daft to do so laboriously by hand what could be done so much faster by machine, and to condemn oneself to economic marginality by refusing to mass-produce things that could be sold in great quantity.

The craftsman may even be ambivalent about his own work. My choosing to design and build a table in the manner of the eighteenth century cabinet-maker John Townsend reflects both my personal esthetic preference for straight lines, restrained ornamentation, and polished plane surfaces, and an undeniably nostalgic longing for an aristocratic lifestyle. To complicate matters further, I increasingly reject the excesses of the lifestyle implied by the table as my politics grow more radical. My awareness of both the elemental needs of the socially dispossessed and the load my materialistic craving places on the ecosphere causes me to say with Thoreau, "Simplify, simplify!" "Thank God," my mentor says, "I can sit and I can stand without the aid of a furniture warehouse."

But even as I find myself reproved by Thoreau, I am aware that in making a beautiful object I am both setting my mind and body against the dehumanizing conditioning supplied me by my own culture — which insists that I be a passive consumer of mass-produced goods — and bringing into being an object that may become a more or less permanent counterpoise on the scales which in our culture are tipped so heavily in favor of the shoddy. Neither the simple esthetic object nor the process of making, I suggest, can be understood apart from their psychological and cultural ground. To attempt to comprehend either is to become aware of how intimately one's sense of self is bound up with what one does (or does not do) with one's hands; of how one's political views interact with one's urge to shape materials into forms that externalize the ideals in one's mind; of how one's freedom to shape materials is conditioned by one's culture, and especially by education, law, and public policy; and of how one's view of the world is some inscrutable function of one's dialogue with materials and forms.

Indeed, thinking about the situation of the craftsman in America inevitably causes one to question many aspects of our common experience, past and present: public and private notions of the self and the social order; images of physical and cosmic reality; systems of value and priority; models of education, production, and consumption; assumptions about creativity; and attitudes toward tools and the human body. Or so I have concluded after having repeatedly been asked four questions about my own work.

1.   Four Questions

Did you take some courses in woodworking?

The simple answer to this question is "No," but more significant than that autobiographical fact is the existence of the widespread conviction that one can do only what one has been formally trained to do. This is one of the enduring ironies of our educational system. A host of critics of American education have recently argued that our schools have the perverse effect of rendering people fearful, indecisive, passive, and psychologically dependent on the very resources originally designed for their liberation. Paradoxically, these negative characteristics seem to *fit* people for socially acceptable lives as worker-consumers.

Whether one agrees or disagrees with Jacques Ellul's dire prediction in *The Technological Society* (1964) about the apocalyptic implications of technology, one of his observations is undoubtedly to the point here: A fundamental aspect of any technological system is the conviction, first stated systematically by the nineteenth-century American engineer Frederick W. Taylor, that there is finally "one best way" to do a job.[2] One implication of this principle for schools charged with supplying workers and managers for American industry is apparent: they exist primarily to teach the one best way. At a deeper level, of course, they *socialize students into believing* that there is always one best way, and that it must be found. A corollary is that anyone who wants to know how to do a job must subject himself to the educational system. From here it is but one step to the assumption behind the question "Did you take any courses?" Anyone who is capable of doing a difficult task well must have had formal instruction in the technique (the one best way) of doing it.

What personal losses result from this assumption? At least the adventuresome spirit of risk-taking; the satisfaction of discovering, through trial and error, that one can solve previously unencountered problems on one's own; the knowledge that indeed there are many good ways to do a job, and that the "one best" (i.e., most efficient) way may not be the most humanly satisfying; and the confidence that one can, as Thoreau insisted, reason from one's hands to one's head. Nor is it necessary to adhere to a romantic conception of human nature to view these personal attributes as either necessary or possible to achieve. Indeed in a broader cultural sense they are vital to the functioning of any complex social order because it requires people who can cope with the myriad of problems that lie outside prior experience, present theory, and established procedure.

The revitalization and humanizing of American culture are contingent upon many things, but among them is certainly the restructuring of the schools in terms of a new understanding of the process of learning. One upon occasion learns formally and in an orderly fashion from someone whose sole relationship to oneself can be described by the term *teacher*. But more often — and more naturally — one learns either through his own purposeful and unaided attempts to solve an immediate problem, or by observing, assisting, and emulating another learner, problem-solver, or worker with whom one has a far more complex relationship: father, mother, friend, colleague, or sibling.

Did you use a kit?

To ask this question is perhaps to admit grudgingly that a degree of self-mastery and manipulative skill remains to some members of the culture, together with a wistful urge to assert oneself against the definer and producer of one's own incompetence. From a certain perspective, it may be possible to take comfort in the enormous increase in kit-making in recent years. My own view is rather less sanguine. Indeed it is closely analogous to Herbert Marcuse's analysis of certain forms of sexual license in a technological culture — forms he calls "repressive desublimation." The idea, as Marcuse outlines it, is that in order to retain the co-operation of its subjects, a technological society allows a certain release of sexual energy, *so long as* it is linked with ultimate fidelity to one's basic role as producer and consumer. Society sanctions the behavior of the Heffnerian playboy only to the extent that he shows himself to need all of the paraphernalia of middle-class consumerism: expensive high fidelity systems and photographic equipment, aquatic gear, clothes, snowmobiles, and so on. The implication is that sexual pleasure is permissible in no other context.[3]

So with kit-building: a sanctioned opportunity to indulge certain impulses, *provided* one does not step outside his authorized socio-economic role as the as-sembler of mass-produced interchangeable parts. To pursue the analogy with Marcuse's repressive desublimation a bit further: the kit-building syndrome not only *allows* certain forms of what must pass for creativity; it also in effect defines what creativity *is*. Such a definition is safe for both the culture and the kit-builder; the culture is not disturbed by bothersome creativity that undermines its indispensable values and assumptions, and the kit-builder can be "creative" without having to first imagine what he will create, face any of the unsettling and even threatening uncertainties the craftsman routinely encounters, or vacate the only role in which he feels comfortable.

Do you have a lot of special tools?

If this question is taken to mean "Do you have more tools than the usual household screwdriver and rusty pliers," then the answer is yes. If it means "Do you have exotic tools especially designed to enable you to accomplish wood-cutting operations that would otherwise be humanly impossible," the answer is no. In either case, the point of the question is that the mere possession of tools is thoughtlessly made equivalent to self-confidence and self-mastery, manipulative competence, and design skill, which are the actual enabling powers. In the view of most people who do not work with their hands, it would appear, tools are repositories of occult power that allows their manipulator a degree of control over physical reality denied to ordinary mortals. Again the corollary is depress-ing: one who does not possess the tools lacks access to the occult power, and there-fore may not respond to an impulse to give shape to material reality.

But tools have only recently come to be viewed in this way. They have tradi-tionally been more the evidence than the cause of competence. Apprentices frequently made their own tools, incised them with their initials, and perhaps even decorated them with engravings, as proof that they had mastered their craft. Thus the tools became evidence not only of self-mastery, but also of the personal vision of reality held by the craftsman — artifacts of his attempt to synthesize inner and outer reality, or vectors indicating the probable future results of his attempt to impose form on the world about him.

If it is also true, as I suspect it usually is, that the image of a tool my interrogator has in mind is a *machine* tool rather than a simple chisel or plane, then the question has further implications. Because "tool" in our culture has come to mean "machine tool" almost by definition, at least two corollaries follow. First, the machine tool is both metaphor and extension not of a single controlling human consciousness but of the impersonal logic and efficiency of a production *system* — that is, it *is* the system in miniature. Only such a system could produce an object. Hence the craftsman is viewed as *one who can operate the mini-system* — one who in his very creativity testifies to its legitimacy and benign helpfulness. Second, one produces an object by means of those operations a machine tool is designed to perform; to understand the design of a whole object is to understand the sequence of special machine tool operations required to produce it. But in reality most craft work — done by hand with hand tools — is of a vastly different character: measure, cut and try, look, feel, turn this and that way to the light, take a little off over there, improvise a tool or technique — and so on through the whole catalogue of essentially improvisatory procedures every craftsman learns to rely upon.

The question reveals ultimately, it seems to me, how fully we have become tools of our tools, wishing fervently not for the patience, self-confidence, and self-mastery required of the craftsman, but for *access* to the mini-system and the expertise required to operate it. The difference between self as *maker* and self as *operator* is psychological as well as functional, as many a proud new owner of a Shop-Smith must have discovered to his sorrow.[4]

Why don't you go into the business?

"Not long since," Thoreau says in *Walden*, "a strolling Indian went to sell baskets . . . in my neighborhood. . . . Having seen his industrious white neighbors so well off . . . he said to himself: I will go into business; I will weave baskets; it is a thing which I can do." Reflecting on the incident in the context of his Walden experiment, Thoreau continues, "I too had woven a kind of basket of a delicate texture, but I had not made it worth any one's while to buy them. Yet not the less, in my case, did I think it worth my while to weave them, and instead of studying how to make it worth men's while to buy my baskets, I studied rather how to avoid the necessity of selling them."

The idea that one can make an object that could find a profitable market, and yet have no desire to sell it, does not spring easily to the minds of many people. Absent from the question is any sense of the complex relationship between maker and object. An object one has made may not be *sale-able,* not psychologically *available* for sale in the usual sense. To sell it might imply disposing of the quantum of self it contains, the hold on reality it has allowed one to attain, and the sense of release involved in seeing what is inside come outside and achieve form — the pleasure of saying "That is me; that is part of what I am inside."

But since many craftsmen sell what they make — indeed make it for sale — Thoreau's little parable does not fully explain the relationship between the craftsman and his work. It is not finally the act of selling itself that distinguishes the craftsman from the assembly-line producer of mass-distribution goods, but the way he understands and approaches selling, and especially his refusal to allow selling to control either his work, with which he strongly identifies, or the pattern and rhythm of his life. Items at craft sales, for example, are customarily signed or marked by the maker, and a few are usually designated "not for sale". One need merely contrast this situation with the counter of any department store — where the items are identical, none are marked by the worker who produced them, and all are most emphatically for sale — to see how far outside the American production-consumption ethos the craftsman stands.[5]

Indeed, the cultural anthropologist Jules Henry has distinguished between primitive and contemporary cultures partially upon this basis. "The primitive workman," Henry notes in *Culture Against Man* (1965), "produces for a known market, and he does not try to expand it or to create new wants by advertising or other forms of salesmanship. . . . [He] is content to be underemployed at his craft if it does not keep him busy. Thus there is a . . . relatively stable relationship not only between production and material needs, but also between production and psychological ones: the craftsman does not try to invent new products . . . nor to convince his customers that they require more or better than they are accustomed to . . . . [Neither] are they moved to come to him with unmet yearnings they must satisfy with new products. The contrast between primitive culture's assumption of a fixed bundle of wants and our culture's assumption of infinite wants is one of the most striking — and fateful — differences between the two cultural types."[6]

Ours is not a primitive culture (at least not in Henry's sense), but the continued presence of craftsmen within it testifies to the historical survival — if not to the economic viability, which I will discuss later — of a peculiar self-understanding and a set of values long since purged from mainstream America. The craftsman's aim is not to make as many items as he can entice people to buy by means fair or foul, but to make each object as an expression of his own self-understanding and view of reality, to keep it if it means enough to him, to sell it if anyone wishes to buy it, and in either case to make a better and more beautiful one next time. The craftsman is not "in business" in the usual sense, even if he offers his wares for sale.

2. Crafts and the Business of America

The business of America is a special kind of business, after all, but it is frequently used unquestioningly as a universal standard against which the validity of every kind of business activity must be judged. Thus crafts have been viewed — at least officially — as either having little to do with larger social, economic, and political questions, or as a poor relation of the "natural" and beneficent urban and industrial social order destined in the grand evolutionary scheme of things to supplant it.

Craftsmen in America have for a variety of reasons been at a disadvantage from the beginning, when it was made difficult — even criminal — for European craftsmen to carry tools, patterns, templates or models to the New World.[7] During the colonial period, American craftsmen were "strongly discouraged by a mercantilistic England," as C. Malcolm Watkins and Ivor Hume have shown in a study of the so-called "poor potter of Yorktown." The colonial governor, aware that officials in England would not be pleased to discover a skillful maker of large quantities of earthenware and salt-glaze pottery, mentioned in his reports only a "nameless poor potter" whose work "is so very inconsiderable, that there has been little less of that Commodity imported since it was Erected, than there was before."[8]

Nor did release from the colonial yoke of England significantly improve the craftsman's lot, for the view that national destiny lay exclusively in the direction of encouraging mass production quickly became orthodox after the Revolution. "The expediency of encouraging manufactures in the United States . . . appears at this time to be pretty generally admitted," Alexander Hamilton said before the House of Representatives in 1791. "A man occupied on a single object [i.e., production operation] will have it more in his power, and will be more naturally led to exert his imagination, in divising methods to facilitate and abridge labor, than if he were perplexed by a variety of . . . dissimilar operations."[9]

Public policy during the ensuing two centuries has remained oriented toward industrialization, and at best neutral on the subject of crafts. Thus the craftsman remains at a disadvantage in our culture, not because of the intrinsic nature of his work but because of public attitudes and policies. Although a Bureau of Labor survey as early as 1904 had located twenty-five significant crafts revival projects in twelve states, its condescending conclusion was that the revival was important "chiefly . . . [to provide] employment for persons living in rural districts and having little else to occupy their time . . . and also for city men and women who are incapable of supporting themselves at more difficult [i.e., industrial] occupations."[10] Even Oscar L. Triggs, who helped established a pioneering arts and crafts league in Chicago around the turn of the century, came to the inexplicably sanguine conclusion that "with the absorption of the higher type of mind in industrial pursuits [exploitative] conditions are destined to pass away."[11] Andrew Carnegie's *Gospel of Wealth,* published a few years earlier in 1899, apparently convinced Triggs that "those who have great wealth by means of the present system are seeking to know their duty in regard to that wealth." The machine and the trust, he fancied, had a beneficent effect: men and women will continue to love to work, and "the machine and the trust will but increase their economy [i.e., efficiency]". In retrospect, such a conclusion is difficult to reconcile either with Triggs' own work with the league, or with his mentor William Morris' philipics against the "tyranny of the excess of the division of labor."[12]

A striking example of the practical results that derive from viewing all crafts activity condescendingly and any industrialization reverentially is the Appalachian crafts revival that began in the 1890's — surely one of the most significant that has ever occurred in America.[13] Although most of the traditional arts and crafts that had long flourished — largely out of necessity — in the Appalachians had disappeared by the late nineteenth century, crafts began to be revived after about 1890. By 1935 there were at least twenty-seven important centers of crafts activity in the region, involving no fewer than 250 chair makers, thirty to forty musical instrument makers, many potters, and innumerable makers of baskets, quilts, and other items.

One might logically assume that so dramatic a revival must have been supported by public laws, policies, and funds. Not so. Its support came almost entirely from Hull House, Berea College, and the home missions boards of several religious denominations. Allen Eaton, who in the 1930's surveyed forty years of craft activity in the Appalachians, reported in 1937 that only marginal support came from public funds and agencies: some sporadic efforts of the Extension Division of the U.S. Department of Agriculture and the Division of Subsistence Homesteads of the Department of the Interior, token funds administered through the Smith-Hughes Vocational Act (1917), and some small programs later connected with TVA. State aid was nearly negligible.[14]

The latter decades of the Appalachian crafts revival coincided, more over, with the coming of good roads, public schools, and especially industry to the region — developments that received massive public assistance and political support. State legislatures refused time after time to impose taxes on industry, either because they feared taxes would retard economic development or because the legislatures themselves were controlled by commercial interests. High school and college curricula were tailored expressly — as they still are — to the needs of the industrial and commercial interests that control the economy of the region.

The results of such a scheme of priorities should have been predictable. The crafts revival reached its peak about 1935; crafts activity in the region has been barely marginal since. And although coal production in Appalachia tripled between 1900 and 1950, almost none of the benefits of such "economic development," as it is still so quaintly called, passed to the people of the region. Eighty-five percent of the coal mined in West Virginia is shipped out of the state, yet the first governor who seriously tried to impose an extraction tax on coal (William Marland, 1953-57) died as a taxi-cab driver in Chicago, ruined by the powerful interests he tried to control. West Virginia retains even yet one of the most regressive tax structures in the nation, yet state and federal courts have repeatedly upheld archaic broad form deeds that encourage mining operators to plunder rich coal-bearing lands they neither own nor pay taxes on.[15]

Amid the ruins of a scale house long since abandoned by a coal operator in the desecrated hills of eastern Kentucky lives a skilled chairmaker who is a consummate example of the human results of warped public policies. Wendell Berry has written of him: "Unable to live by his work, [he] is dependent on the government's welfare program, the benefits of which are somewhat questionable, since if he sells any of his work his welfare payments are diminished accordingly, and so he stands little or no chance of improving his situation by his own effort. Only the workman's loving pride in his work can explain why he has continued to make any effort at all....It is not possible to escape the irony of the fact that ... a man of skill and industry, whose craft is itself one of the valuable resources of his region and nation, and who . . . [makes] products of great beauty and usefulness — is destitute in America, now."[16] What he needs are tools; what he gets are food stamps that in the end cost more than tools. Yet the same government that denies the very legitimacy of his work and thereby condemns him to live as a scavenger in an industrial wasteland, bows time after time to the efforts of the coal lobby to defeat any and all legislation calculated to guarantee the necessities of life for the region's people.

3. Constructing a Counterargument: the Failure of Conventional Wisdom

Until recently, those who sought to challenge public policies in the interest of placing the craftsman on a firmer footing have been dismissed as cranks, reactionaries, and nostalgic dreamers who could neither adjust to the hard realities of progress nor comprehend the marvels of the free enterprise system. But they have recently begun to find allies in quite unexpected quarters.

It is true that if one does not question the "growth myth" on which modern technological development is based, crafts work is not economically viable: it is too slow and costly. But ecologists and other political and social critics have so thoroughly discredited the logic of an open-ended growth policy that the whole question of mass production vs. crafts may have to be reopened. And with it, the matter of public policy with respect to crafts. Manufacturers speak glibly of producing items at "low cost," but we are increasingly aware that the calculated cost of production is perhaps only a small fraction of the total social cost. The total cost — including environmental damage, substandard wages, the depletion of resources that by rights should belong to the public, and public (tax) subsidies to private industry — may be very near what the item would have cost if produced by hand.

Mass production is now "necessary", after all, partly because we have been conditioned to "need" goods in great quantity and variety. The imperative for mass production arises out of a peculiar set of values that is itself partially the result of the mass-production system. Like the highway trust fund, it is a self-generating system that lacks any ability to criticize or check its own growth. But to the extent that the values, assumptions, and human implications of our system of production and consumption are challenged, our artificially created and sustained needs will surely begin to erode, and with them part of the foundation upon which our mass-production ethic rests.

But how does one answer those who argue that industrial development has achieved such momentum that one has no alternative but to "adjust" to its human and social implications? Rene Dubos, pioneer microbiologist and investigator of human adaptation, has argued eloquently that "Adjusting man's philosophical perspective to modern technology seems . . . at best a dangerous enterprise," for the simple reason that such adjustment frequently denies man's ineradicable

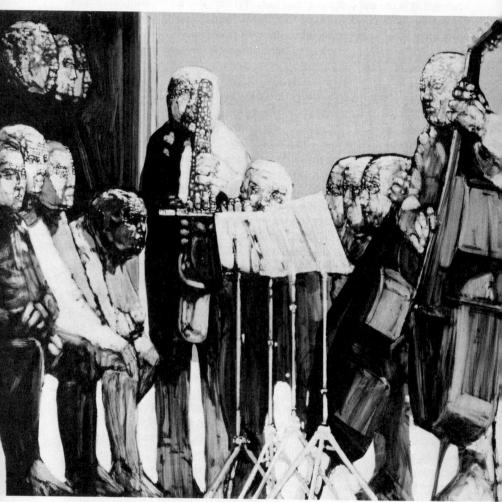

"Musicians", 3' x 5' (Acrylic).

needs for sensory engagement with physical reality, of which he is deprived by life in an urban-industrial environment. "Whatever scientific technology may create," Dubos argues, "*l'homme moyen sensuel* will continue to live by his senses and to perceive the world through them. As a result, he will eventually reject excessive abstraction and mechanization in order to establish direct contact with the natural forces from which he derives the awareness of his own existence and to which he owes his very sense of being."[17]

Those who wish to dismiss the cultural importance and viability of crafts are thus obliged to construct new arguments. But at the same time, those who comprehend the personal and cultural value of crafts work must argue more intelligently in its behalf. The argument that will convince the unconvinced and bring about changes in public policy is yet to be constructed, both because many of its components are not yet known and because there is as yet no willing audience for it. To suggest seriously that crafts can figure importantly in cultural renewal is to encounter at least incredulity if not skepticism and hostility. Conventional wisdom (especially that used to justify public policy) is settled upon the subject: crafts are intrinsically marginal, anachronistic, non-competitive, merely quaint, and so on through a whole catalog of fashionable condescension. We must extrapolate all current development curves or die.

Luckily, the actual dynamics of history frequently do not follow the scenarios so confidently prepared by purveyors of conventional wisdom. During the past decade, for instance, in the face of all the realistic, mature advice they have been receiving through both media and major institutions — "Plastics!" Benjamin is advised by his self-appointed mentor in *The Graduate* — the young have turned to crafts in increasing numbers.[18] Their work may well reflect some of the critical shifts in consciousness, values, and behavior that will ultimately be necessary if crafts are to make their full potential contribution to the process of cultural renewal. Finding their inspiration in the work of contemporary anthropologists, ecologists, and radical political theorists, young craftsmen seem to view their work not as escapist and nostalgic, but as a variety of political activism: to reduce one's wants and thus to curtail one's consumption of the plastic products of the system is a revolutionary act since our system depends upon high rates of consumption. With an admirable logical consistency, young craftsmen also shun the synthetic materials of American industry — vinyls, for example — in favor of organic materials: wood, stone, clay, leather, natural yarns, and so on. Indeed, they reserve the term "plastic" to denote anything artificial, phony, or lacking in integrity and durability.

But more than an admiring view of the counterculture and a predilection for radical politics will be required if we are to improve the situation of the craftsman — both for his own benefit and for the enrichment of American life. Against the conventional wisdom that is ossified in current policies we must erect a counterargument and propose a new set of policies, which take into account both the stubborn realities of our culture and the available points of leverage within it.

The counterargument must rest upon more than simplistic economic considerations; it must, for example, take into account the capacity of crafts work to define and sustain identity, and its valuable non-alienated and non-alienating character. It must emphasize its quite unexpected tendency to allow interventions of the sacred and transcendent into mundane reality, thereby enriching private experience and investing one's world with meaning.

In his explorations of the growth of individual identity — a "subjective sense of invigorating sameness and continuity" — Erik Erikson discovers a correlation (particularly in preadolescence) between identity and the confidence that one can make things: "a developing sense of industry." All children, he says, "sooner or later become dissatisfied . . . without a sense of being able to make things and make them well and even perfectly. . . ."[19] Unfortunately, our culture frustrates this developing sense in a thousand subtle ways. Our schools are oriented toward the production of high verbal and analytical, but low manipulative, ability. Children are surrounded from nursery school onward by polished, complicated objects they cannot hope to make, their workings — as a colleague once put it — "hidden by the industrial designer's slickest art." Hence their development of a sense of industry is either truncated or never begun. Moreover, since teachers in the public schools are traditionally women, and *knowing* is separated from *making* and *doing* (which are masculine), children have great difficulty achieving a crucial sense of identification with 'those who know things *and* know how to do things," most of whom are *men*, outside the schools. In view of such a lamentable situation, the cultural value of the craftsman lies in his sense of competent industry — his confidence that he can both manipulate materials and shape his own identity, and his constant awareness that each implies the other.

Even more importantly, Erikson notes the "unity of personal and cultural identity." When we try to sketch the development of identity, he says, we "deal with a process 'located' *in the core of the individual* and yet also *in the core of his communal culture,* a process which establishes, in fact, the identity of these two identities." Thus an object made by a craftsman is not the result of an "inner" imagination and creativity; it is not a "private" contribution to "culture"; it does not take its place as one more inert item in our "external" world. If Erikson is correct, such separations are false. The object expresses the essential unity of inner and outer worlds, the craftsman's attempt to integrate them even more completely, and the cherished inseparability of individual and cultural identity.

Implicit in Erikson's argument about work and identity is another point which one must go to Norman O. Brown and Herbert Marcuse to find elaborated: since identity is a partial function of one's sense of one's own body, and because the "sense of industry" Erikson speaks of is related to manipulative skill (i.e., control over the body), work must reinvigorate the body if it is to help create and sustain identity. Brown was not talking about crafts, certainly, when he began to use the phrase "the resurrection of the body" in a special sense.[20] The change in the self that many craftsmen (particularly young craftsmen) are seeking depends, in any case, upon a rediscovery, reintegration, and reinvigoration of the senses — a veritable resurrection of the body. The choice of organic materials as a means of reestablishing contact with a primal reality is paralleled by an impulse to ground one's being in renewed sensory awareness and vitality. The fact seems to be that Dubos is right: the body has an irreducible need for sensory stimulation, which most work that is available to us does not provide.

Work that resurrects the body and reintegrates self and world is almost by definition non-alienating, and non-alienating work is difficult to come by in our culture. But craftsmen have long known through their own intuition and experience those truths about work which political and social critics like Marcuse derive from Freudian and Marxian theory. Work in our culture, Marcuse says in *Eros and Civilization* (1955), is usually "such that the individual, in working, does *not* satisfy *his* own impulses, needs, and faculties . . . ." Such work is "alienated labor" — necessitated by a perversely symbiotic combination of a represssive ego and a performance-oriented society. But non-alienated *work* is finally inseparable from bodily pleasure: "If pleasure is indeed in the act of working and not extraneous to it, such pleasure must be derived from . . . the body itself, and must eroticize . . . the body as a whole."[21]

The connections between sexuality, work, and identity are not commonly recognized in our culture — not only because Brown, Marcuse, Erikson, and Dubos rarely turn up on the coffee tables of middle America, but also because to admit such connections may be deeply threatening to both self and social order. The political and social values, models of self, and views of work and sexuality upon which crafts work is based are fundamentally opposed to most of the determinants that give American society its present shape.

My final conceptual contribution to the counterargument is a frankly speculative attempt to describe a tantalizingly elusive but unquestionably important aspect of the work of at least some craftsmen: they obviously seek a practical political alternative to urban, capitalist, technocratic America, but in a more basic religious sense (and herein lies the speculation) they seek to transcend through their work the finite limitations of perception and knowledge that bind them personally and culturally. Nor are they the first people in history to do so. They stand, as a matter of fact, in a long line of those who have experienced the sacred potential of making things with one's hands.

A Yakut proverb says that "The first smith, the first shaman, and the first potter were blood brothers." Mircea Eliade quotes the proverb in *The Forge and the Crucible: the Origins and Structures of Alchemy* (1962) as he traces, through mythology, the "spiritual adventures" of primitive miners, smiths, and metalworkers who found themselves "aware of their power to change the mode of being of substances."[22] He finds, of course, that a great distance separates us as kit-builders or assembly line workers from primitive *homo faber*, to whom both the cosmos and human existence were sacred, and for whom work had a sacred value. Primitive man could "immerse himself in the sacred by his own work . . .

and as a creator and manipulator of tools." By contrast, men have become "incapable of experiencing the sacred in [their] dealings with matter," since matter itself is viewed as a totally desacralized "natural phenomenon". Eliade's argument can be roughly paraphrased as follows: If the cosmos is perceived as sacred, then changing the mode of being of substances is a sacred activity. And since power over substances implies a knowledge of sacred, cosmic reality, a loss of the sense of cosmic sacrality entails a loss of the sacred significance of making.

"A very old dream of *homo faber*," Eliade says, is that by "collaborat[ing] in the perfecting of matter [man will] at the same time [secure] perfection for himself." In a curious way, this perception is not far from Erikson's argument about the inseparability of identity and culture — of inner and outer worlds. Thus from our particular historical perspective, the one-directional process Eliade describes may in fact be reversible. If viewing the cosmos as sacred sacralizes work, it may also be that work which is (for whatever reasons) meaningful or sacred may resacralize or reinvest with meaning the culture (or cosmos) one lives in. Any activity that carries such a potential within it deserves more than the schizophrenic combination of benign neglect and hostility that has thus far been its fate to receive in our culture.

4. Toward A New Policy

A young craftsman of my acquaintance, standing amid the skillfully made leather goods he offers for sale, recently told me that he was able to obtain no assistance whatever when he attempted to open his shop. Banks would not lend to him because he had neither collateral nor certificate of competence from a college or university, and his work itself — although clearly of high quality — was not admissible evidence. Government loan programs were set up to handle only amounts many times larger than the few thousand dollars he needed. Local public institutions have yet even to recognize that such needs exist.

As a society, we refuse to admit the personal, economic, or social value of what thousands of craftsmen are doing, while at the same time we offer massive subsidies to individuals and corporations — Lockheed, General Dynamics, Georgia Pacific, Peabody Coal, Dow Chemical — engaged in the pillaging, paving, and polluting of America. Wendell Berry's chairmaker, sitting amid the ruins of some coal company's abandoned scale house, is thus no special example of the problems of a single region; he is a veritable archetype of the results of the educational, fiscal, and social policies of our culture at every level.

But by what practical means could the situation be altered? We tend to assume that there are none — that however humanly desirable it would be to encourage crafts work as a source of livelihood for a significant number of people, it "just can't be done" because the whole "current of modern life" makes it "uncompetitive". But it is utterly meaningless to talk about the "competitiveness" of any endeavor without assessing the reasons for its competitive status. We seem caught in an uncritically adopted evolutionary model of the social and political order. We assume that our society has its present shape as a result of glacially paced evolutionary developments that were and are beyond our conscious control. But the fact is that we *shaped* it ourselves by making certain conscious choices. And although some things cannot be changed — "They Can't Put It Back," Mike Kline's song says of the strip-mined hills — some *can*. We *can* make other choices.

A federal agency comparable to the Small Business Administration could be set up to provide some of the same types of assistance to craftsmen that we now provide so freely to those who want to open filling stations, bowling alleys, motels, and franchised doughnut shops.[23]

Small, long-term, low-interest federal loans could be made available to craftsmen for training, materials, and equipment. The demonstration of a skill should be accepted as collateral.

To place the craftsman on a somewhat more equal footing with the oil companies, a "personal energy depletion allowance" could be instituted. Because a craftsman must depend upon his own physical, intellectual, and emotional energy to earn his livelihood, he could be allowed to compensate for the depletion of that energy by having part of his earnings exempt from taxes. Or alternatively,

To make the craftsman more competitive with wealthy individuals *all* of whose income from municipal bonds is tax-exempt, all earnings from the sale of crafts could be tax-exempt.

The federal work-study program could be extended to embrace apprenticeships with established craftsmen. At present, the federal government will pay eighty percent of the wages of a student hired part-time by a university professor or in the library. But a student who wishes to work part-time as an apprentice to a local craftsman cannot usually receive such support because colleges apply most work-study funds to student assistants on campus.[24]

All of these suggestions — and many more are easily conceivable — are completely practical, and could be implemented within present legal, governmental, and economic structures. They presuppose neither political nor social revolution. They would require, however, that we revise our all too automatic assumptions about work, and about the basic organization and aims of our collective cultural and economic life.

That the craftsman is not in business *in the same way,* nor for the same reasons (to say nothing of on the same scale) that Howard Johnson's or IBM or even the local merchant should be beyond question, for example. Yet so otherwise admirable a document as the recent report of the Interagency Crafts Committee asserts that "a crafts project should be evaluated like [presumably meaning "by the same criteria as"] any other business enterprise," and appears to place crafts activity at the bottom end of an evolutionary scale whose top end is "modern industry."

Discussing the development of the craft of hand tufting in northern Georgia, for example, the report says that "Occasionally a home industry of this kind *will develop into a full-fledged machine operation*. . .[this one] has *progressed to pro-ducing* machine-made chenile products and has now *evolved into a sophisticated enterprise.*"[25] Or to take another example from the same report, the reasons for the present low status of crafts and the craftsman are located primarily (though not exclusively) in the craftsman himself: he lacks sophistication, leadership ability, design talent, and entrepreneurial skill. Discriminatory federal tax and subsidy policies receive no consideration, and the loading of the educational system and the media in favor of mass production and consumption enter only inferentially through the observation that crafts activity in the schools has traditionally been slighted.

In effect, then, the Interagency Crafts Committee Report asks merely for a bolstering of what any reasonable person must admit is and will remain a marginal aspect of our national life. The basic economy must remain oriented as it is at present; our cultural life (including crafts) will be "supported" by agency programs, foundation grants, special committees, and the like—be dealt with, that is, as the tertiary phenomenon it is. We agree, it seems, that the thing is to "help" Wendell Berry's chairmaker, rather than to redress the elemental skewing of the system that bequeathed him his dilapidated scale-house dwelling.

But this is a schizoid position, and must be recognized as such. We cannot continue to assume that our "basic economic system" ("basic" is a crucial *metaphor* here, not a mere adjective) must remain as it is *and* that certain cosmetic adjustments can be made in order to offer a few people a mode of life almost totally at variance with every motive, value and assumption upon which that system rests. Nothing is, nor should be recognized as, more basic, after all, than the elemental human needs out of which—partly *in reaction to* our economic system and its manifold cultural consequences—the current crafts revival has arisen. If we mean to deal seriously with the value and policy implications of the current revival, we might well begin by admitting how profoundly it calls that system into question.

REFERENCES

1. It should be borne in mind that I use the term "craftsman" in less than its broadest possible sense. I do not mean to include, for example, craftsmen in the building trades (carpenters, steel riggers) or in manufacturing industries (tool and die makers, machinists), who have a secure position in the American economic system. Not do I include craftsmen (usually middleclass and university trained) who teach in crafts programs in colleges and universities. Credentialed, protected by tenure and guaranteed a salary, they also have been integrated (or co-opted, depending upon one's point of view) into the system. I am interested primarily, then, in the craftsman who has taught himself the skills of his craft or learned them from another working craftsman, who supports (or

attempts to support) himself by his work, and whose choice has placed him—for reasons
I shall discuss—outside the guarantees and supportive structures available to those
willing to accept the more conventional options offered by our society.

An earlier version of this paper was published in *Centennial Review,* vol. 17 (Summer,
1973), pp. 215-236. Portions reprinted here by permission.

2. Jacques Ellul, *The Technological Society,* trans. by John Wilkinson (New York: Vintage, 1964). The principle is also discussed extensively by Daniel Bell in *Work and Its
Discontents: the Cult of Efficiency in America* (Boston: Beacon Press, 1956). The full
psychological, social and political implications of the general acceptance of this principle are too extensive to discuss here.

3. Herbert Marcuse, *One-Dimensional Man* (Boston: Beacon Press, 1964) and *Eros and
Civilization* (Boston: Beacon Press, 1955). Any issue of *Playboy* will provide abundant
documentation for this assertion.

4. A Shop-Smith is a special convertible mini-system designed for the woodworking
hobbyist.

5. A friend who was for years a shop-steward in the automobile industry told me of assembly-line workers who periodically marred an auto chassis or body with a hammer out of
frustration at having no legitimate way to record the fact that it had been worked on by
an identifiable human being—no way to say, *"I* worked on *this* automobile."

6. Jules Henry, *Culture Against Man* (New York: Vintage, 1965), p. 9.

7. Archie Green, *Only a Miner: Studies in Recorded Coal Mining Songs* (Urbana: University of Illinois Press, 1972), p. 113.

8. C. Malcolm Watkins and Ivor N. Hume, *The "Poor Potter" of Yorktown* (Washington,
D.C.: Smithsonian Institution Press, 1967), pp. 75-79.

9. Alexander Hamilton, "Manufactures," a Report to the U.S. House of Representatives,
December 5, 1791. *The Works of Alexander Hamilton,* ed. Henry Cabot Lodge (New
York: G. P. Pubnam's Sons, 1903), vol. IV, p. 70 and p. 88.

10. Max West, "The Revival of Handicrafts in America," *Bulletin of the U. S. Bureau of
Labor,* Vol. 9 (November, 1904), p. 1596

11. Oscar Lovell Triggs, *Chapters in the History of the Arts and Crafts Movement*
(Chicago: Bohemia Guild, 1902), pp. 184-186. An excellent survey of the arts and crafts
movement in this period may be found in Robert J. Clark (ed.), *The Arts and Crafts
Movement in America, 1876-1916* (Princeton: Princeton University Press, 1972).

12. See William Morris, "The Arts and Crafts of Today," in *The Collected Works of
William Morris* (London: Longmans, Green, 1914), vol. XXII, pp. 356-374, and "The
Revival of Handicrafts," *ibid.,* pp. 331-341.

13. My account of the Appalachian revival draws heavily upon Allen H. Eaton, *Handicrafts
of the Southern Highlands* (New York: Russell Sage Foundation, 1937).

14. See Eaton, *Handicrafts,* pp. 77 and 251-299. The Smith-Hughes Vocational Education
Bill (1917) provided for education in agriculture, trades and industries.

15. William Blizzard, "West Virginia Wonderland," *Appalachian South,* Vol. 2 (Spring-
Summer, 1966), pp. 9-16.

16. Wendell Berry, *The Long-Legged House* (New York: Audubon Ballentine, 1971), pp. 3-
11.

17. Rene Dubos, *So Human an Animal* (New York: Scribner's, 1968), pp. 196 and 20.

18. This salient feature of the counterculture has received little comment outside the
pages of the *Whole Earth Catalog,* ed. Stewart Brand (Menlo Park, Calif.: Portola Institute, 1968-71). William Braden, *The Age of Aquarius* (Chicago: Quadrangle Books,
1970) skirts it altogether. Lew Yablonsky's *The Hippie Trip* (New York: Pegasus, 1968)
mentions it briefly (p. 306), as does Naomi Fergelson's *The Underground Revolution*
(New York: Funk and Wagnalls, 1970), p. 173. Theodore Roszak's widely read *The
Making of a Counterculture* (Garden City, N.Y.: Doubleday, 1969) alludes to it several
times, but never discusses it.

19. Erik Erikson, *Identity: Youth and Crisis* (New York: W. W. Norton, 1968), pp. 19, 123ff.

20. Norman O. Brown, *Life Against Death* (Middletown, Conn.: Wesleyan University Press, 1959).

21. Herbert Marcuse, *Eros and Civilization* (1955; rpt. New York: Vintage, 1962), p. 201.

22. Mircea Eliade, *The Forge and the Crucible: the Origins and Structures of Alchemy.* Trans. by Stephen Corrin (1956; rpt. New York: Harper and Row, 1971), p. 7. Eliade's analysis of the sacred dimensions of myth is too extensive to detail here. Its essential points are synthesized in Mircea Eliade, *The Sacred and the Profane: the Nature of Religion,* trans. by Willard R. Trask. (1959; rpt. New York: Harper and Row, 1961). A very brief synopsis may be found in my earlier article, "The Sacred and the Profane in *Walden," Centennial Review,* Vol. 14 (Summer, 1970), pp. 267-283.

23. A promising bill to create a National Folklife Center (S. 1844) has recently been introduced in the 93rd Congress by Senator Abourezk (D.-S.D.) and others. The National Endowment for the Arts, in cooperation with the National Council on the Arts, has belatedly (as of 1972) established some programs for encouraging craftsmen and supporting crafts activity, but available grants are small and restrictive. Limited assistance to craftsmen has recently been made available also through two agencies of the U. S. Department of Agriculture (the Farmer Cooperative Service and the Extension Service). The Economic Development Administration of the Department of Commerce assisted with some crafts projects in the late sixties. For a brief survey of the efforts of federal departments and agencies, see Charles Counts, *Encouraging American Craftsmen: Report of the Interagency Crafts Committee* (Washington: Government Printing Office, 1971), pp. 27-30. To my knowledge, this report is the best and most complete practical discussion of the problem of crafts policy now in print. My comments on the report below are intended not as criticism, but rather as contributions to an argument that most urgently needs to be made, and toward the formulation of which the work of Counts and others is an admirable contribution. The Interagency Crafts Committee is made up of representatives of government departments and agencies.

24. Sargent Shriver once testified before the House Committee on Education and Labor's subcommittee on the War on Poverty that "there is no advantage in restricting the administration of work-study programs exclusively to universities," but such a restriction seems to have been followed in practice, nevertheless.

25. Counts, *Encouraging American Craftsmen,* pp. 6-8. The tufted carpet industry of northern Georgia (e.g., Cabin Crafts, Inc.) bears not the remotest resemblance—economically, culturally or (one might add) politically—to the craft of hand tufting once practiced by lone craftsmen in the mountains.

Art, Technology, and Ideology: The Bauhaus as Technocratic Metaphor

John Adkins Richardson
Professor of Art and Design
Southern Illinois University

How to understand aright the kind of unity history imposes upon a milieu is a difficult and puzzling thing. Rarely can we do justice both to the nature of the society as a whole and to the character of the individuals who compose it. That a unity *does* obtain seems certain. For individuals—no matter how gifted—realize themselves only in society and develop their geniuses only through relations with others. It is, therefore, common to speak of an age or generation in terms of a consensus of ends, means, purposes and ideals. But that is only one side of the matter. There is a corollary characteristic: no individual is ever wholly absorbed by his culture and no ideas can be wholly accounted for by it. There is a sense in which men and their works remain invincibly insulated, and synopses invariably do violence to the elusive and endlessly varied particulars of existence. Still, in the absence of synoptic thought, history is an empty chronicle and its practitioners are apt to lose their ways in a tangle of unrelated causes. Moreover, a general view can sometimes reveal a genuine commonality among particulars where diversity and antagonism would seem to prevail. The discovery of such affinities is essential to examination of the role of institutions in society. The way in which a synoptic view can overcome real but misleadingly apparent differences can be demonstrated in no better way than by looking at the Bauhaus in its time and place.[1]

The Bauhaus was a famous school of industrial design established in Germany in 1919 by architect Walter Gropius. In 1925 it moved from the city of Weimar to Dassau where it continued until 1933 when it was closed by the German government. Universally, it is held to have been the most influential institution of its type. The goal of the Bauhaus was to reintegrate the creative artist into the reality of modern industrial society and, simultaneously, to humanize the exclusively material attitudes of the business community.[2] More than any other single influence the Bauhaus brought together under one rubric the diverse strands we now call "modern art." Several men connected directly or indirectly with German Expressionist, Dadaist, and Surrealist movements were on the staff. And its principal spokesman, Laszlo Moholy-Nagy, was unremitting in his praise of the avant-garde antirationalists. Commenting on the rejection of Surrealism by the Soviets, he wrote: "Again and again artists must state that revolution is indivisible and that the intellectual and political strategy of the revolution must be accompanied by a long term emotional education. Only correlation and integration can bring a change in habits and attitudes of the people rooted in and grown out of previous conditions."[3] Elsewhere, he noted a connection between this prescription and the work of Freud:

> Through the use of this psychological insight and the psychoanalysis of Sigmund Freud, space-time fundamentals may be considered also as the syntax and grammar of emotional language which may re-create the path of the inner motion. This can express the problems of living (through the arts) more directly and synchronously in their totality than could be done by any mere descriptive version. And as people have to learn to read and relate the manifold signs of traffic control at metropolitan street intersections for physical safety, they must learn also to read and relate the emotional meaning of the expressive fundamentals used in the different arts in order to avoid the danger zones of psychological "intersections."[4]

It would be unfair to equate the explicit philosophy of any painter with Moholy-Nagy's view. He spoke for himself and not for those whose work he admired. But his rather touching faith in the priority of artistic sensibility over political radicalism, his conviction that a particular mental set will sustain individual freedom within a highly organized society is very close to the posture of the German Expressionists prior to the war of 1914. Their freedom, like that of professors, was altogether "academic." What Expressionism, and also Dada, Surrealism, and even the Bauhaus credo come down to is celebration of the subjective, the intimate, and the personal, and the rejection as an illusion of the objective existence of a collective body termed "the masses." Socially, they hoped for a more active and individualistic conception of human relationships but their rebellions against middle-class ethical and aesthetic values—an opposition constantly proclaimed—had nothing whatever to do with opposition to the political realities of bourgeois life. Quite the opposite in fact.

Since 1911 the idea of freedom as a condition of art and culture has become increasingly powerful in the modern world. It is reflected in the popularity of such devices as the phonograph which enables one to exercise freedom of choice in the privacy of one's chambers simply because one does not have to listen to a program of selections determined by some concert master. Painting, in particular, is seen as an opportunity for the uninhibited expression of strong, even antisocial feelings and a consequent experience of utter freedom, within the bounds of grace and generalized attractiveness.[5] In some quarters the conviction persists that release through art can beneficially affect one's social behavior, that the particular discipline of the arts carries within the seeds of humane propriety.[6] That was very much the German Expressionist attitude. When Wassily Kandinsky, for example,

spoke of liberty it was almost always in terms of artistry where, he said, there is "a great freedom which appears unlimited to some and which makes the *spirit* audible."[7] It is to the point that in his *Reminiscences,* speaking of the student uprisings in Moscow in 1885, he remarked: "Luckily politics did not completely ensnare me. Different studies gave me practice in 'abstract' thinking, in learning to penetrate into fundamental questions."[8]

The sort of obsession with personal liberty that Kandinsky and his followers sought sustains the middle classes. It does not really matter whether the "wants" that keep the system functioning are for bread, serious literature, personal services, orange juice, or marijuana. What is important is the preservation of private decision, particularly where those decisions relate to autonomous cultural activities—the arts, chess-playing, spontaneous socializing.

Fascist and communist aestheticians are perfectly correct when they assail modern art as "bourgeois". Not only is it true in etiological terms, it is so in its entire philosophy. It was inevitable that Expressionism, Dada, Surrealism, and Bauhaus thought—the products of an elite underground intellectual group—should eventually be domesticated by the blithe sophisticates who populate the allied worlds of Fashion and High Society. Their appetites for novelty are so rapacious that they will consume the most pungent items as though these were delicacies solely for their delectation. From digestion by the aristocrats of bourgeois taste it is but a single step to popularity. The validity of this assumption is proved by the prevalence of Bauhaus inspired furnishings in suburbia, the ubiquity of slab-shaped glass skyscrapers in New York, Berlin, and London, and (most of all) the frequent use in advertising of Mies van der Rohe's "Barcelona Chair" of 1929 as an ensign of contemporary good taste.

At the root of anything the Bauhaus did was "functionalism," the theory that aesthetic content will virtually take care of itself if a thing is designed to fulfill its purposes with maximum utility, ignoring all precedents and habits of construction. Students in the school were given bizarre problems that challenged their imaginations and forced them to undertake empirical experiments with basic materials. They made springs of wood, structural supports of folded paper, designed better-than-igloo domiciles for Eskimos, created paintings to "modulate" light rather than merely reflect it. They played with illusion as they did with stress, and with psychic effects as with die-casting, and all in the name of function. It was not until after World War II that a discipline called "engineering psychology" rose into prominence, applying research into anatomy, perception, and time-motion to work methods, environments, and equipment design, but Bauhaus experiments concerning the effectiveness of different shapes, sizes, and materials anticipated almost all later research into tool and furniture design.

One must be rather cautious here, however, for it is well known that Bauhaus philosophy opposed the meretricious uses to which businesses put psychology. Despite the fact that many of the most skillful art directors in advertising agencies are products of Bauhaus oriented curricula it is true that the intentions of the staff were quite noble. M.I.T.'s Gyorgy Kepes echoed Bauhaus convictions when he wrote: "If social conditions allow advertising to serve messages that are justified in the deepest and broadest social sense, advertising art could contribute effectively in preparing the way for a positive popular art, an art reaching everybody and understood by everyone."[9]

Besides, it can be shown that the aims of Bauhaus researches into perception were either aesthetic, directed towards creation of a new iconography, or were purely functional. An example of the first kind of interest has been provided by Hirschfeld-Mack, a member of the school throughout its existence:

> A very interesting seminar was held during those early years. It was under the leadership of Paul Klee and Wassily Kandinsky and others. They sought to discover the reactions of individuals to certain proportions, linear and color compositions. . . . In order to find out whether there is a universal law of psychological relationship between form and color, we sent out about a thousand postcards to a cross section of the community asking them to fill in three elementary shapes, the triangle, square, and circle with three primary colors, red, yellow, and blue, using one color only for each shape. The result was an overwhelming majority for yellow in the triangle, red in the square, and blue in the circle. This was only one of the many problems which was tackled in this seminar.[10]

That such discoveries can have commercial application was not the reason for seeking them, but it is not hard to see that application of them to practical ends was promoted by the orientation of the curriculum.

The inclination of the school program was quite consistent with the historical situation that had brought it into prominence. For Bauhaus ideals mirrored German hopes between the two world wars. They reflected the idealism of the Weimar Republic, whose optimism in liberal reform and technological progress shows too in many social doctrines popular at that time. The Republic itself hardly had a chance, saddled as it was with all the sins of the defeated Empire. But that was by no means apparent in the twenties.

After the establishment of the Dawes Plan of reparations payments (which began by lending Germany 200 million dollars), and the conclusion of the Locarno Agreements stabilizing the frontiers, a new period of prosperity opened for the Germans. Municipalities and businesses obtained huge loans from the United States to renovate factories and re-establish the merchant marine. "Made in Germany" became again a familiar notice in the world of commerce. Reparation payments were made on schedule from 1924 through 1930. The application of reason to political circumstance seemed promising indeed.

But German intellectuals were new at having anything to say about the operations of representative government. Before the war they had been free and active with respect to the spiritual and artistic spheres but passive and even apologetic about the political realm of human affairs. Indeed, many of them scorned such interests as being profane. Conditioned by life in a land where everyone wanted to be known as *Herr Doktor, Herr Professor, Herr Regierungsrat,* or what have you, rather than by such comparatively democratic titles as France's "Citizen," the "Mister" of the U.S., or the "Comrade" of Russia many non-Communists arrived at the conclusion that a technocracy—a government run by experts—would be a fine solution to the major problems of society. Radiant faith in the ability of "designers" to run the world shines through everything Moholy-Nagy wrote. Heidelberg University's Professor Karl Mannheim was of a similar faith. But he had thought through many of the political problems of modernity. To treat them he invented what he called "the sociology of knowledge."

In his most famous single work, *Ideologie und Utopie,* published in 1929, Mannheim began with the recognition that disinterested intelligence, represented by the intelligentsia, had been disengaged from practical affairs for at least a century. The objective of the sociology of knowledge was to recapture tnat intelligence and anchor it in social institutions.[11] Mannheim wished to fulfill the liberal ideal of applying organized intelligence to human affairs. But he also recognized that human nature tended to fault dreams of social tranquility rather than support them. Before him Marx had argued that most men saw the world from a distorted angle because their world views were created by social institutions instead of the other way around. According to Mannheim, Marx had not gone far enough. Mannheim had read Eduard von Hartmann[12] and Sigmund Freud as well as Marx and it shows clearly in the development of his own theory. He wrote:

> The sociology of knowledge is concerned not so much with distortions due to a deliberate effort to deceive as with the varying ways in which objects present themselves to the subject according to the differences in social settings. Thus, mental structures are inevitably differently formed in different social and historical settings.[13]

Mannheim was saying, then, that whole social classes of society, not just individuals, are controlled by the unconscious. All social thought, no matter how ostensibly "objective," represents a particular viewpoint because every man occupies a specific social habitat and his experience brings him into contact with only a limited range of experiences. Inevitably, every opinion is partial and there is no such thing as an objective, unbiased truth.[14]

Up to this point the sociology of knowledge may seem little more than discreetly stated iconoclasm, dedicated to shattering the intellectual foundations of government and replacing politics with anarchy. That is not the case however. For, according to Mannheim, the absence of final truth is but one of two realities with which modern society must contend. The other is a need for social control. Human beings prefer anything to chaos; in modern times dictatorships are the usual response to social disorganization; they represent the coercive rationalism of society. Men threatened by anarchy will almost always accept the imposition of arbitrary "truths" simply for the sake of order.

What Mannheim wanted instead of either chaos or dictatorship was a planned but unregimented society, a marriage of social control and bourgeois freedom. To achieve this end he proposed a practical strategy of social reform. He argue that, while all social classes are caste-marked by partly closed minds, there is one class in European society that has a special immunity to narrowness. Historically, it is a subclass, that part of the bourgeoisie to which Mannheim himself belonged, the intelligentsia. The chief characteristic of this group is that it has been drawn increasingly from other than the middle classes and is, in any case, but loosely anchored in the outlook of any class, since, "Due to the absence of a social organization of their own, the intellectuals have allowed those ways of thinking and experiencing to get a hearing which openly competed with one another in the larger world of the other strata."[15] He was eager to get this class, whose members could break through the rigid ideologies that separate other men, into society's "loci of decision."

Since Mannheim had no desire to disenfranchise anyone, he came to the conclusion that it is "the task of the political leader deliberately to reinforce those forces the dynamics of which seem to move in the direction desired by him, and to turn in his own direction or at least render impotent those which seem to be to his disadvantage."[16] Every successful politician does in fact behave according to this precept, of course, but in Mannheim it is something more than a description of necessity. Speaking of consumer goods and merchandising, he writes that a "planned but unregimented society" should aim at "inducing conformity not by authority or inculcation, but by skillful guidance, which allows the individual every opportunity for making his own decisions. A planned society would direct investment, and by means of efficient advertisement, statistically controlled, would do everything in its power to guide the consumer's choice towards coordinated production."[17]

Because he held that it is only by remaking man himself that it is possible to reconstruct society along these benevolent patterns, it should not surprise us to learn that he, like Bertrand Russell and Moholy-Nagy, was an admirer of John Dewey's philosophy of education.[18] American readers may feel that this is anything *but* the way to produce intellectuals, but the learned of nations having a long classical tradition are less enthralled by their systems of public and private education than are Americans. For them the irrelevance of disengaged, cultivated intelligence to the domain of citizenship has been proven over and over and over again. Indeed, it is that very detachment Mannheim hoped to overcome with his sociology of knowledge. What interested him about the intelligentsia was not the brilliance or erudition of its members but its encompassing social outlook. In consequence, the education of leaders would not be intelllctually competitive so much as psychologically reinforcing and conducive to social awareness.

In like manner, the Bauhaus program (which extended from kindergarten through the university level) was what we today would call "Progressive Education." Moholy-Nagy maintained: "Teaching focussed on learning for learning's sake will always bypass the ultimate objective which alone can give sense to the attempts of integration. Choices are easy if the goal is clear. Knowledge should not be suspended in a vacuum; it must be in relationship with socio-biological aims. This integration gives to human life content, direction, and a sense of security."[19] Moholy-Nagy uses Julian Huxley's *Evolution: The Modern Synthesis* to argue for recognizing that in human life as in the rest of nature the results of natural selection and the adaptations it promotes are by no means necessarily "good" from the point of view of either the species or of the progressive evolution of life. According to his line of argument, it would be both foolish and wicked for an economy or an educational system to copy the methods of so imperfect a system. Of course, the difficulties with this argument from analogy with modern theories of evolution are identical with those entailed in justifying *laissez-faire* economics with Spencerian or Darwinian principles, but it will serve no useful purpose to debate educational or social theory here.

The point is that these Weimar theorists felt that the future of mankind lay in the development of individual capacities. Such an approach to education would, they believed, insure modern society of a necessary supply of specialists while simultaneously providing it with a benevolent and responsive leadership. Such confidence in the self presupposed, if not Expressionism, at least the general state of mind with which Expressionism was identified.

When Bauhaus spokesmen talked of design being an "attitude more than a profession" they were propounding a rationalized version of expressionist will. Moreover, these industrial designers and the Expressionists had more in common than merely an overlap of some members of the Bauhaus teaching staff. Among the German Expressionists as in the Bauhaus and the Sociology of Knowledge the idea of some new communal existence for individuals was important. And all three groups put their faith in strong, radical methods; they shared a dynamic outlook on life and there was a common tendency to reject the past except where it served specific immediate purposes. Expressionism presumed a vast indefinite nature that enveloped and engulfed the individual; Moholy-Nagy spoke of "organic totality, and Mannheim of the "collective unconscious." Although the prewar Expressionists were emotional and ecstatic whereas the Weimar intellectuals appeared cool and calculating there was a peculiar affinity between the groups and sometimes, as in the cases of Kandinsky, Klee, and Feininger, an actual overlap.

Against this view one might contend that, given the exaggerated statement of the parallels, one could bring even the Nazis under the same roof as the Bauhaus and Karl Mannheim. That is perfectly true. But grouping the Nazis with their despised enemies is not so preposterous as it may sound. The mere fact that Hitler drove the Bauhaus and Mannheim into exile along with every other cosmopolitan does not mean that he and they have no common ground.

It would, of course, be monstrous to contend that the likenesses that may obtain between the Nazis and the expressionist-technocrats are much more than conjectural ones. In at least one respect their world views have nothing whatever in common. It is at the very point in which Expressionists join with bourgeois ideals — the point at which both hold as a primary value the liberty of the intimate. That particular kind of freedom was what Freud,[20] the Expressionists, the Dadaists, Surrealists, and the Bauhaus wished to attain, regardless of the costs. And it was for *this* reason the Nazis distrusted all of them. Hannah Arendt has put it sharply and, I think, correctly:

> If totalitarianism takes its own claim seriously, it must come to the point where it has "to finish once and for all with the neutrality of chess," that is, with the autonomous existence of any activity whatsoever. The lovers of "chess for the sake of chess," aptly compared by their liquidator with the lovers of "art for art's sake," are not yet absolutely atomized elements in a mass society whose completely heterogeneous uniformity is one of the primary conditions for totalitarianism. . . . Himmler quite aptly defined the SS member as the new type of man who under no circumstances will ever do "a thing for its own sake."[21]

No one can maintain that the Bauhaus and Nazism are allied in any but the most coincidental ways. Regimentation was a primary condition of the National Socialist community. Its methods were radical in the sense of being inhumane but they were not particularly innovative — unless one counts the creation of an entire technology for mass murder. The vast indefiniteness that engulfed the Nazi was racial, the blood consciousness of the German soil and the *Herrenvolk*. If the Bauhaus represents Expressionism rationalized then the Third Reich must represent it gone insane. In practice they have little in common but an interest in machine design and a taste for clean and regulated posters. Nazi oppressiveness is far from the unregimented society described in *Ideology and Utopia*. Still, if one reads Mannheim's prescription for the planning of history he will see that it would not have been out of place in the pages of *Mein Kampf*. Mannheim said that, in the planning of the historical process, actions were not to be directed primarily to attaining "the best possible qualities or to following the most favourable path, but towards the means which are most likely to lead from the status quo to the desired goal and which gradually transform the person who uses them."[22]

We can go further still, remarking just one of several weaknesses in Mannheim's educational theory.[23] That is that the intelligentsia as Mannheim finds it is a wildflower, existing by virtue of *post*-educational classlessness. It could very well be that the group is not subject to domestication. Certainly, it is odd for him to suppose that one can have an egalitarian system and still support a specialized elite whose purpose is the planning of history. If the dynamics of leadership are to be so institutionalized someone must at some time initiate the "chosen" into elitehood and it must be made perfectly clear to them that they constitute a distinctly separate order of person. The whole procedure is reminiscent of the techniques used to train and select the leadership from among the German youth.

The resemblances between Nazism and Expressionism are sufficiently striking to have once inspired an attack upon the art style by the famous Marxist philosopher and literary critic, Georg Lukacs. In 1934 his article "Grandeur and Decay of Expressionism" appeared in the journal *International Literature*.[24] He attempted to prove that Expressionism was nothing more than a preparation for fascism and was, in fact, intimately bound up with it. He contended that Expressionism (which he tends to identify with all modernist trends) deals with the world as if it were no more than the product of individual psyches and helps to promote the view that history is dependent upon a collective something called the "national psyche" which in turn is a fundamental of fascist thought.[25] Two years later Lukacs was forced to retract this argument because of complaints from Expressionists active in underground Communist activities in Germany. It is not easy to estimate the meaning of the retraction, however, because Lukacs has always been a controversial figure in Communist circles and in 1933 he had already repudiated as "bourgeois idealism" everything he had written, including *History and Class Consciousness*, which had had a considerable influence on Mannheim. By the 1950's these "ins and outs" had taken on the character of a farce since by that time Lukac's attitudes on Expressionism had become the official Party Line.

What Lukacs had correctly estimated in Expressionism was its lack of any wholeness of vision. The movement was not just unorthodox and antitraditional, it was unrealistic in a verifiable way; that is, its partialisms were *intentionally* partial. Ecstatic in its acceptance of individual separateness, it deified the autonomy of artistic expression. A committed leftist would not be deceived by the pretended antibourgeois spirit of the avant-garde; Lukacs realized that Expressionism bore a direct relation to the structure of the society from which it sprang.

His error, or so it seems to me, lay in mistaking for a single limb the intertwined branches of an enormous tree. But standing as he was within the very shade of that tree it was not easy to see the details. He was even able to confirm his interpretation by citing a letter that Goebbels had written to Eduard Munch, complimenting his later works. Lukacs could actually have bolstered this line of argument by reference to Emil Nolde, a prominent German Expressionist who had been an original member of the Nazi Party in North Schleswig. Nolde's autobiography, *Jahre der Kampfe* (1934), was intensely anti-Semitic and his art sometimes gives vent to the same prejudice. When twenty-six of his works were included in the 1937 exhibition of "Degenerate Art" he wrote letters of protest to the Ministry of Culture and Dr. Goebbels, trying to convince them that his style was consistent with Aryan principles.[26] It was to no avail. In 1941 the government forbade him to paint at all, even as an avocation. Nolde's is one of history's oddest examples of unrequited political love. But, of course, it does not prove that Expressionists are spiritual relations of fascists. Nolde had made the same mistake as Lukacs, taking separate branches for a limb.

Those branches, Expressionism and fascism, were in contact at certain points and they were sustained by the same root system, but they grew from different boughs. To extend the metaphor we may compare Expressionism and the Bauhaus with branches which spring in different directions from one bough. Growing quite close by on the Bauhaus side is another offshoot, the sociology of knowledge. And, inevitably, there are subsidiary twigs and dead sticks woven into a maze of overlappings. All are sustained by the peculiarly modern hope of attaining truth by bringing disparate subjectivities into some sort of collective whole. This hope has been one of the most conspicuous elements in the intellectual climate of the nineteenth and twentieth centuries.

As early as 1912 we find Kandinsky saying: "Just as each individual artist has to make his word known, so does each people, and consequently, also that people to which this artist belongs. This connection is mirrored in the *national element* in the work."[27] Even Mannheim's notion that intellectuals can give us an overall perspective "that takes account of all partial perspectives is a reflection of the popular notion that truth is discovered by listening tolerantly to all points of view."[28] The Bauhaus combined these conceptions: if a sense of community and openness does not arise spontaneously, it can be acquired through technical adjustments of the means of communication and production. The Nazis were never content to rule by state power and its machinery of violence; like the Bauhaus liberals they desired men voluntarily to behave in ways to support the society. Hitler's psychology of terror is no more than the darker side of the technocratic coin. Not only totalitarian political propaganda but all advertising and modern mass publicity play on fear to some extent and contain an element of threat.

Social and political theorists have not failed to remark the general paradox of which our topic is but a special case, namely the appeal of mass movements for highly individualistic, cultivated people. The affiliation of a Nolde with Nazism, Degas with the anti-Dreyfusards, Pound with the fascists, Sartre with the Communists is sometimes explained away as another evidence of neurotic self-hatred and compensatory selflessness. At other times such behavior is interpreted as the arrogance of an elite who hope to control others by association with a movement. Actually, as Hannah Arendt has noted, "the much-slandered intellectuals were only the most illustrative example and the most articulate spokesmen for a much more general phenomenon."[29] The modern mass movements were preceded by periods of extreme individualization in Germany, Italy, Russia, even the Orient. They attracted into the fold typical "non-joiners" who were adrift in an atomized society long before and much more easily than they did the more sociable members of traditional political parties. Thus, it is not really surprising to find Nolde committed to Hitler; the early Party was almost exclusively composed of misfits and remained to the very end dominated by what Konrad Heiden very aptly called "the armed bohemians."[30] Mannheim and the Bauhaus

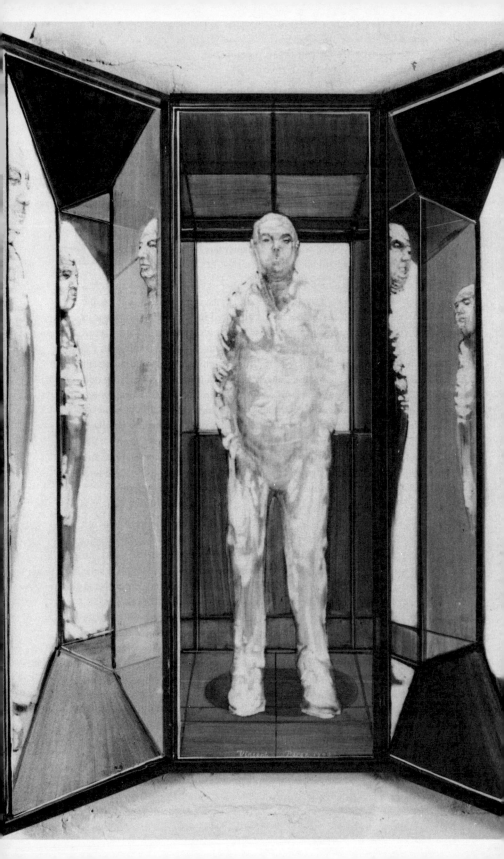

staff recognized that the competitive structure of modern society had as a concomitant feature a sense of anomie and isolation that could no longer be held in check by membership in the traditional social classes. They sought to effect a new sense of community through humanistic liberalism. But their gradualisms were no match for a movement which treated all moral precepts as pious banalities.

What is interesting about the parallels between Expressionism, the Bauhaus, and the Nazis is that they reveal just how closely contained is everything within a given social framework. One consequence of this is that every intellectual activity in a given time involves comparisons with bordering activities, not because of any ideological relationships but because the problems men are predisposed to solve in an age are very much the same.

We who live today are struggling with the same sense of hopelessness that the Weimar intellectuals and the Nazis hoped to surmount. So far none of our institutions — not state, school, church, family, community, or eccentric subculture — has been able to overcome it. And it is perfectly within keeping with this situation that we should be hearing Mannheim-like paeans to educational reform and social planning. Too, just as before, many people who deprecate respected standards and minimize accepted theories do so because they feel imperiled by institutional rigidity. Some social critics find it extremely odd that these same individuals should take counsel from an inventive iconoclast like Buckminster Fuller. In his book *Generations: A Collage on Youthcult* Clifford Adelman complained:

> Fuller's persistent judgment that the reformation due is to the environment and not to man himself, removes — at the hands of those who are so inclined — the responsibility for inner change. It is not surprising, then, that Fuller himself explains away all sins, even those on the order of Mylai or Kent State, in terms of "inertia," i.e., they are morally neutral and the inner self never has to answer for them. From that it is even less shocking that Fuller's World Game winds up a cosmic fascism: We surrender ourselves to super-visionary design-scientists who run the spaceship while we sit back and "do our thing." The kids who are trying to get ahead of that game now, citing Fuller as a resource, will move us toward that state faster than they realize.[31]

Whether the prototypical experience of the Weimar Republic is to any avail as prediction is, of course, uncertain. Although its brief history holds little consolation, the one absolutely sure thing about the past is that it is foredained never to be repeated in more than the most general sense. Apart from that consideration, our examination of the time has value if only to the extent that it demonstrates that the coincidence of Bauhaus ideology with Nazism and the Sociology of Knowledge is not merely accidental.

REFERENCES

1. Much of the following material is derived from Chapter VII of my book *Modern Art and Scientific Thought* (Urbana, 1971).

2. Cf. Laszlo Moholy-Nagy, *Vision in Motion* (Chicago, 1947), pp. 10-12, 63-112. This book is primarily concerned with the "American Bauhaus," the Institute of Design in Chicago, but is a very complete representation of Bauhaus educational policy which, as the author notes on page 84, was absorbed and furthered at Chicago.

3. Moholy-Nagy, p. 340.

4. Moholy-Nagy, p. 115.

5. See, for example, Meyer Schapiro, "The Liberating Quality of Avant-Garde Art," *Art News,* Vol. 56, no. 4 (Summer, 1957), pp. 36-42.

6. For example, John Dewey, *Art as Experience* (New York, 1934) and Herbert Read, *Education Through Art* (London, 1943) argue this.

7. Wassily Kandinsky, "On the Problem of Form," trans. Kenneth Lindsay in *Theories of Modern Art,* ed. Herschel B. Chipp (Berkeley, 1969), p. 159.

8. Wassily Kandinsky, *Reminiscences,* trans. Hilla Rebay and Robert L. Herbert in *Modern Artists on Art,* ed. Robert L. Herbert (Englewood Cliffs, N.J., 1964), pp. 24-25.

9. Gyorgy Kepes, *Language of Vision* (Chicago, 1947), p. 221.

10. Ludwig Hirschfeld-Mack, *The Bauhaus* (Croydon, Australia, 1963), p. 6.

11. Charles Frankel, *The Case for Modern Man* (New York, 1955), p. 122.

12. Hartmann was the first to devise a complete doctrine of the Unconscious. See Eduard von Hartmann, *The Philosophy of the Unconscious,* trans. William Chatterton Coupland (London, 1893).

13. Karl Mannheim, *Ideology and Utopia,* trans. Louis Wirth and Edward A. Shils (New York, 1957), p. 265.

14. Cf. Frankel, pp. 122-26.

15. Mannheim, p. 12.

16. Mannheim, *Man and Society in an Age of Reconstruction* (New York, 1940), p. 247.

17. Mannheim, *Man and Society in an Age of Reconstruction,* p. 315.

18. In America Mannheim's principal spokesmen have been educational theorist George Counts and the Social Reconstructionists.

19. Moholy-Nagy, p. 24.

20. See Sigmund Freud, *Civilization and its Discontents,* trans. Joan Riviere (London, 1939), *passim.*

21. Hannah Arendt, *The Origins of Totalitarianism* (New York, 1951), p. 322.

22. Mannheim, *Man and Society in an Age of Reconstruction,* p. 223.

23. More important, but less easy to develop briefly, is the objection that Mannheim misunderstands the role of objectivity in science and confuses the whole issue of the distinction between the study of human behavior and the study of nature. The Frankel book has a very good section on this (pp. 129-33). The most questionable part of the whole theory is the feasibility of planning history for a society that is not regimented. Friedrich A. Hayek in his *The Road to Serfdom* and Julian Benda in *La Trahison des Clercs* have brought devastating criticism against the idea. (Neither of these books is as illiberal as many leftists pretend and many rightists hope.)

24. See Georg Lukacs, "Gross und Verfall des Expressionismus," *Probleme des Realismus* (Berlin, 1955), pp. 146-83.

25. Fear that subjective-idealism and similar attitudes of mind will lead to improper lines of thought is a characteristic anxiety of old line Communists. The most embarrassing example is Nikolai Lenin's polemic against Ernst Mach's philosophy of Empirico-Criticism in *Materialism and Empirico-Criticism,* the only extensive piece on Marxist

philosophy undertaken by Lenin. It represents a complete misunderstanding of Mach's real position.

26. See Paul Ortwin Rave, *Kunstdiktatur im Dritten Reich* (Hamburg, 1949), p. 74.
27. Kandinsky, "On the Problem of Form," p. 157.
28. Frankel, p. 129.
29. Arendt, p. 316.
30. See Konrad Heiden, *Der Fuehrer: Hitler's Rise to Power* (Boston, 1944), p. 100.
31. Clifford Adelman, *Generations: A Collage on Youthcult* (New York, 1972), p. 71.

Institutions
Where Technology
and Human Values
Meet:
An Inventory

Belinda Barrington

Institutional Programs on Science and Society.
Subcommittee on Science, Research, and Development, Committee on Science and Astronautics, U.S. House of Representatives, 92nd Congress, 2d Session, "Teaching and Research in the Field of Science Policy — A Survey," Committee Print, 20 December 1972, 110 pp. Listings of course titles and names of instructors at a number of colleges and universities.

Socio-Engineering at Southern California.
University of Southern California, Socio-Engineering Program, School of Engineering, University of Southern California, University Park, Los Angeles, California 90007. A new undergraduate engineering curriculum is being developed at the University of Southern California, designed to stress the interrelationships among psychology, sociology, political science, law, economics, fine arts, and technology. An important aspect of the program will be the "encounter" — active involvement by the student in actual technological, social, and political problems and processes. The "encounter" will take the form of interaction with community groups and visits to the community, coupled with lectures, discussions, and seminars. The senior year curriculum will include an Engineering Semester, in which the students participate in actual engineering projects in the Los Angeles area. The program director is the dean of the School, Dr. Zohrab Kaprielian.

Technology and Social Policy at Berkeley.
University of California, Berkeley, Science, Technology, and Public Affairs Project, Berkeley, California 94720. The Science, Technology and Public Affairs Project at the University of California at Berkeley is an informal, cross-disciplinary program of instruction and related research that spans formal and informal undergraduate courses and faculty and graduate seminars. An Advisory Committee on Science, Technology and Public Affairs has been set up to facilitate the

development of programs that cross discipline and degree lines. Parallel research efforts are being carried out by the Lawrence Radiation Laboratory and the Institutes of Transportation Engineering, Urban and Regional Development, International Studies, and Governmental Studies. A description of the Program and an initial list of courses may be obtained from Dr. Todd LaPorte, Department of Political Science.

Colorado Center for Science and Politics.
University of Colorado, The Edward Teller Center for Science, Technology, and Political Thought, c/o Dr . Edward J. Rozek, P.O. Box 2336, Boulder, Colorado 80302. The Edward Teller Center for Science, Technology, and Political Thought was incorporated in July 1973 at the University of Colorado. The Center proposes to sponsor educational programs and conferences to interest students, academicians and the general public in the role technology plays in solving social problems and protecting human freedoms. The Center is expected to engage in research and instruction in science, technology, and the science of government and international politics. The Center will also study the "interaction between science, technology, and politics in order to find a way to maximize man's ability to control physical forces unleashed by science and technology."

Interschool Program at Washington University.
School of Engineering and Applied Sciences, Washington University, Program in Technology and Human Affairs, Washington University, Box 1106, St. Louis, Missouri 63130. The Program in Technology and Human Affairs offers degrees at both the undergraduate and graduate levels. The undergraduate program emphasizes a study of the social sciences and experience at the technology-society interface, using project work and policy studies to bring the students into direct interaction with public and private agencies and community groups. Areas of activity include technology and education, international development, health care, and popular education. The graduate program allows students to focus their technical skills on problems of housing, education, communications, environmental quality, and urban, rural, and international development. The program includes work in technology assessment and public policy decision-making. It is intended, at both graduate and undergraduate levels, to address the problems that emerge from the interaction of technology and society, and to progress towards solutions of those problems. It has close ties with the Center for Development Technology, at Washington University. Inquiries may be addressed to the program chairman, Dr. Robert P. Morgan.

Interdisciplinary Center at Washington University.
Washington University, Center for Development Technology, Box 1106, Washington University, St. Louis, Missouri 63130. The Center for Development Technology is a research, development, and training center at Washington University. The work of the Center spans a broad range of science and technology policy; issues include assessment of the economic, environmental, and social impacts of

new technological devices, methods, and materials, the application of technologies to international socio-economic development, and the application of existing technologies to meet domestic needs for communication, education, energy, and housing. The Center maintains close ties with the Technology and Human Affairs Program at Washington University as well as with counterpart institutions overseas to coordinate the Center's work in international development. The Center's research and development activities provide training for students of many disciplines, and its staff members are involved in the development and teaching of interdisciplinary courses on the interaction of technology and society. Program information and a publications list may be acquired from the director, Dr. Robert P. Morgan.

University of Wisconsin Program in Technology-Values.
University of Wisconsin at Milwaukee, Cultural and Technological Studies, Systems-Design Department, College of Engineering and Applied Science, University of Wisconsin, Milwaukee, Wisconsin 53201. The Cultural and Technological Studies Program of the University of Wisconsin, Milwaukee, is an interdepartmental effort of the College of Letters and Science, the College of Engineering and Applied Science, and the School of Architecture. Engineering and liberal arts professors cooperate to develop undergraduate course sequences that consider the humanistic implications of technological solutions to national problems. The program is designed to correct a "low interest in people-oriented values" common, the program directors feel, to engineering students. The program consists of an initial battery of "gateway courses" that combine a conceptualization of social values and technical processes. The student goes on to a more intensive study of a particular "cultural paradigm" and concludes with a senior seminar directed at identifying the social responsibility of the professional in contemporary society. A program description is available from the director, Dr. Raymond Merritt.

Engineer-Government Liaison Group.
Engineers Joint Council, Technology Assessment Panel. c/o Dr. Robert Mayer, A.T. Kearney Associates, 100 S. Wacker Drive, Chicago, Illinois, Chairman. The Engineers Joint Council has recently established a Technology Assessment Panel to coordinate the efforts of various professional engineering societies in the area of technology assessment. The panel will provide technical consultation to the legislative and executive branches of the Federal Government in their technology assessment efforts, and will serve as a liaison for a working relationship between the Office of Technology Assessment and industrial and professional engineering groups.

Special Interdisciplinary Studies at Case Western Reserve University.
Case Western Reserve University, Humanities and Social Science Program for Case Undergraduates. Division of Special Interdisciplinary Studies, Case Institute of Technology, Case Western Reserve University, Cleveland, Ohio 44106. The

Case Institute Program is a relatively traditional program of manditory humanities electives for all undergraduates. Three of the electives must constitute a sequence, which is intended to "encourage students to explore in depth a field in the humanities or social sciences, to gain an appreciation of its complexities . . . and to experiment with the concepts and techniques of analysis developed to study them." The final course in the core sequence is an interdisciplinary Senior Symposium that brings together students who have pursued different fields of study, for consideration of timely topics, such as changing concepts of utopia, and the role of the U.S. in the Third World. Some of the possible course sequences are interdepartmental, such as the sequences in environmental policy and geography. Other series relate more directly to the technical disciplines, including the sequences in history of science and science, technology, and public policy.

A Social Direction for Engineering at Vanderbilt.
Vanderbilt University, the Socio-Engineering Program, Vanderbilt University School of Engineering, Nashville, Tennessee 37235. Socio-engineers, in Vanderbilt's terms, are professionals who are liberally educated and socially responsible as well as technically skilled. "The socio-engineer will be equipped to devise valid solutions to complex sociological/economic/political/technological problems. He will understand social processes and be skilled in the methodology needed to plan, operate, and control large-scale socio-technological systems." Vanderbilt is in the process of redirecting its engineering department to produce such socio-engineers. The effort is massive, involving the retraining of faculty members and the restructuring of the School along an intersecting matrix pattern of administrative divisions and academic programs. This matrix arrangement facilitates the development of interdisciplinary projects and increases program flexibility. New undergraduate courses are being developed, a graduate program in socio-engineering is planned, and a new National Center for Socio-engineering is planned, which will sponsor major research efforts on societal problems. Dr. Frank Parker, of the School of Engineering, is the Program's Director.

Bioethics.
The Joseph and Rose Kennedy Institute for the Study of Human Reproduction and Bioethics, Georgetown University, Washington, D.C. 20007. The Kennedy biology and medicine, and to facilitate informed professional and public discussion of issues. Institute members participate in an interdisciplinary program combining clinical care, empirical research, and theoretical analysis of biomedical issues, and sponsor a number of courses in bioethics at Georgetown University. The Institute is composed of the Laboratories for Reproductive Biology, the Center for Population Research, and the Center for Bioethics. The life scientists at the Laboratories for Reproductive Biology and the demographers at the Center for Population Research study the qualitative and quantitative aspects of human reproduction, such as congenital defects, contraceptive devices, and fertility rates. The legal, religious, and philosophical scholars at the Center for Bioethics treat such subjects as human experimentation, genetic manipulation, and death and

dying. Other Institute activities include the preparation of a series of television programs on bioethics for the National Broadcasting Company, an *Encyclopedia of Medical Ethics,* and the development of an interdisciplinary ethics library and *Bibliography in Bioethics.*

Innovation and Public Policy Goals.
National Science Foundation, National Research and Development Assessment Program, Washington, D.C. The National Research and Development Program of the National Science Foundation was established in August 1972 for the purpose of providing "an analytical capability, consonant with the needs of Congress and the Executive Office of the President, for objective study and assessment of how science and technology contribute to the achievement of national goals and objectives." The program must develop new structures for the systematic study of innovations, considering their net benefits to society. One program element will examine past, present, and alternative future policies and practices of governments and assess the relationship between policy options and the process of technological change. One element will examine the socio-economic consequences of technological innovations, and establish methodologies for further study. One element will be concerned with gaining a fuller understanding of the process of technological innovation, and identifying the screens and incentives that operate in the decision process. A final aspect of the program will be the establishment of a data and information support system, including the identification of data gaps and the development of ways to overcome them.

The National Endowment for the Humanities and the National Science Foundation.
Program on Ethical and Human Value Implications of Science and Technology, National Endowment for the Humanities, Washington, D.C. 20506, and National Science Foundation, Washington, D.C. 20550. The National Science Foundation and the National Endowment for the Humanities are collaborating on a new program that will support research and related studies on the ethical and human value implications of science and technology. The program focuses on cultural and humanistic values as they are challenged, modified, or influenced by advances in science and technology, and touches on, but is not limited to, the concerns of technology assessment, environmental impact, ethics, moral philsoophy, and law. Proposals for research may be submitted to either sponsoring organization, depending on whether their primary orientation is in the humanities or science area. Inquiries and proposals should be directed to the Program of Science, Technology and Human Values in the Office of Planning of the National Endowment, or the Ethical and Human Value Implications Program of the National Science Foundation.

New Technology Incentives Program.
National Bureau of Standards, Experimental Incentives Program, Administration Building, A-724, Washington, D.C. 20234. The Experimental Technology Incen-

tives Program has been established within the National Bureau of Standards to obtain knowledge and experience concerning the process of technological invention and innovation in the private sector in the U.S., and to determine how the Federal Government can best increase the rate at which new technologies are successfully introduced and commercialized. The program assumes that technological invention and innovation in the private sector is significantly influenced by federal policies and programs, and that introduction of new technologies increases U.S. productivity, improves the U.S. competitive position, and increases America's capability to solve domestic problems. The program will direct experiments and analyses in three policy areas: possible new approaches to government procurement practices as an incentive to technological invention and innovation; the use of federal regulatory powers to stimulate technological innovation; and special provisions to help individual inventors and small firms to stimulate technological invention and innovation. These investigations will be conducted in cooperation with technically-oriented private organizations and other federal agencies, working closely with the related program in the National Science Foundation. Descriptive material and information regarding participation in the program may be obtained from Dr. F. Karl Willenbrock, Director.

Cultural Studies and Civil Engineering at Princeton University.
Princeton University, Humanities in Civil Engineering, Department of Civil and Geological Engineering, Princeton University, Princeton, New Jersey 08540. Since 1962, the civil engineering faculty at Princeton has pioneered a program linking engineering and architecture, emphasizing the integration of cultural values, history, and technology. The program combines the elements of criticism (in the engineering analysis of impressive contemporary buildings), biography (in the study of structural pioneers), and history (in the study of buildings significant to the architecture of the past.) The humanities are taught by trained engineering faculty, not by humanities faculty. Present buildings are studied as expression of human ideals; past buildings are studied as crystallizations of the lines of technical development and the values of an age; and structures from all periods are related to developments in the visual arts. Among the structures studied have been the Staten Island Ferry Terminal, the Brooklyn Bridge, contemporary works of civil engineering in the Netherlands, and the gothic Cathedral of Rouen; case studies of all of these structures have been published by the program, and may be obtained from Professor David P. Billington, of the Department of Civil and Geological Engineering.

Princeton Conferences on Humanistic Studies in Engineering.
Princeton University, Civil Engineering: History, Heritage, and the Humanities, David P. Billington, School of Engineering, Princeton University, Princeton, New Jersey 08540, from whom conference reports may be obtained. The first Conference, held October 14-16, 1970, dealt with the relation of civil engineering to cultural themes. The second Conference, held in October 1972, also at Princeton,

commemorated the life of Robert Maillart. The Conference sponsored an exhibition of art based on Maillart's bridges, produced a volume of background papers, and climaxed with a series of presentations on the meaning of Maillart's example for contemporary civil engineering practice and education. By encouraging the study of the lives and work of such landmark figures in engineering and design, the Conference organizers hope to enable young engineers to see their own professional roles in a wider humanistic context.

The Man-Made World: Developing Technological Literacy.
Engineering Concepts Curriculum Project, ECCP Regional Center, Room W245, Montclair State College, Upper Montclair, New Jersey 07043. The Man-Made World Program is a laboratory science course intended for senior high school students. It is designed to introduce students to the complexity of technological problems and solutions without imposing on them the formality of traditional technical instruction, and to heighten their awareness of the interaction between technology and social, economic, and political problems. The program format begins with a "state of the art" study of our current technology, and proceeds to a discussion of real-world problems, with a look at information-systems science and engineering as an approach to problem-solving. The *ECCP Newsletter* serves as a clearinghouse for information about the programs at the various schools involved in the Man-Made World Project, and provides technical notes intended to supplement texts provided by the Project. Other available materials include a teacher's manual which offers open-ended guidelines for project initiation, and an explanatory brochure published by McGraw-Hill.

Program for General and Continuing Education in the Humanities at Columbia.
Columbia University, University Directions II, Columbia University, New York, New York 10027. University Directions II is a plan for expanding and reorienting the humanities curriculum at Columbia, guided by the university's awareness that the study of the humanities "spreads outward from a human center and the humane scholar continues to be concerned with questions of value in the midst of increasingly specialized studies on the frontiers of advancing knowledge." The humanities curriculum is being extended from the College to the professional and graduate schools, and interdisciplinary course offerings are being strengthened by interdepartmental and interschool collaboration. The program does not limit its concern to the role of the humanities in technical education; but the program institutionalizes a deep appreciation of the value of humanistic studies, and will explore alternative conceptions of humanity and the humanities, especially as they are seen in other cultural traditions.

National Centers for Study of Values and Environment.
Faith-Man-Nature Group, P.O. Box 397, South Coventry, Connecticut 06238. The Faith-Man-Nature Group was founded in 1965 to draw together natural and social scientists, philosophers, theologians, and generalists seeking to "identify, appreciate, and creatively interrelate the basic values which underlie personal, social and global environmental attitudes, behavior, and decision- making." The

group has held several national working conferences, operates three regional workshops, and has published two books and an occasional newsletter.

Science, Technology and Society at Cornell University.

Program on Science, Technology, and Society, 632 Clark Hall, Cornell University, Ithaca, New York 14850. The Cornell Program on Science, Technology, and Society was established in 1969 "to stimulate interdisciplinary activities in teaching, research, and public information involving the interaction of modern science and technology with traditional social and political institutions and the effect of these interactions of the quality of individual lives." Topics of concern to the program include the ecological impact of technology, public policy for the support and development of science and technology, the impact of technology on values and processes of socialization, and the legal and moral implications of biomedical practices. A substantial amount of program time is devoted to the development of interdisciplinary research, and a number of case studies, papers and legislative testimonies have emerged directly from program activities. Two related Cornell programs are the Program on Peace Studies and the Program on Policies for Science and Technology in Developing Nations. A complete program description and course list, an evaluative progress report, and a publications list are available from the program director, Dr. Franklin A. Long. The Cornell program is compiling a handbook of science, technology, and society programs throughout the country, including non-academic institutions and foundations, serial publications, and information about grants and fellowships. The handbook will be published in the fall of 1973.

Rensselaer Polytechnic Institute.

"Building Educational Bridges between Science and the Humanities," 10 pp, 22 Mar. 71. Troy, New York 12181. The Rensselaer Project has developed a course suggested for use in the final two years of the undergraduate curriculum. The program is designed to integrate the physical, biological, and social sciences with the humanities and fine arts, thus providing a basis for dialogue among students having varied professional interests. A set of guidelines for the course may be obtained from Dr. V.L. Parsegian, Rensselaer Professor, at the Institute. The guidelines propose five course themes dealing with evolutionary processes in nature and society; the long-range viability of human society, the historical and social significance of science and technology, various aspects of communication in natural and social processes, and a comparison of basic concepts of science and the humanities. The guidelines further provide criteria with which to evaluate course material intended to develop these themes.

International Conference and Communication Center.

The Institute on Man and Science, Rensselaerville, New York 12147. The Institute on Man and Science is an educational center, chartered by the regents of the State University of New York, The center sponsors and conducts research, seminar, and special demonstration projects on selected social problems of "an evolving technological society." Past themes have included historic restoration, the United

Nations and the population problem, man and the media, household technology and the quality of life, and a special summer program comparing the key problems of fifth century Athens and modern America. The Institute enrolls members from all professions and publishes a quarterly newsletter as well as interpretive reports of its various programs. Information about the Institute may be obtained from Gordon A. Enk, Director of Research.

Worcester Polytechnic Institute: A Plan for Personal Growth.
Worcester Polytechnic Institute, The W.P.I. Plan, Worcester, Massachusetts 01609. The W.P.I. Plan, as it is called, is Worcester Polytechnic Institute's attempt to provide a humanistic and technical education for its students. It goes beyond other such plans, however, in its concern for the development of human "ingenuity." The plan was developed in 1968 in a climate of concern for the direction of technical education in America, summarized by one W.P.I. professor: "Until we get true perspective on science as a human activity through study of all other human activities, we won't even be good scientists." The plan provides the student with the opportunity to design his own education. Learning modes include the traditional lecture, seminar, tutorial, and independent study, as well as extensive off-campus programs. The academic calendar is unusually flexible and work is individually-paced and problem-oriented. One element of the plan is the Integrative Studies Program, in which students plan and carry out projects integrating the sciences and humanities; one such project studies the historic parallels among periods of great scientific discovery. The stated rationale behind the program goes to the heart of a concern for the incorporation of humanistic values in technology: "Creativity in science springs from the same source as in art or anything else, ... the mind of man and his best work weren't compartmentalized, but a sort of continuum."

Concourse: A "Teaching Community" at M.I.T.
Massachusetts Institute of Technology, The Concourse Program, Massachusetts Institute of Technology, Cambridge, Massachusetts 02139. The Concourse Program at M.I.T. began in 1970 as an effort to achieve "collaborative teaching" among a small group of faculty and graduate students of varied professional interests. The program is open to undergraduates during their first two years at M.I.T., and is directed towards a "natural synthesis and contrast of humane and technical disciplines". The program consists of General Meetings, which involve faculty and students in a single integrated study theme that runs throughout the academic year, and Working Groups, in which small groups of students delineate, plan, and carry out projects. The program is intended not only to encourage the integration of humane and technical values in the prospective engineer, but also to develop the student's ability to work in groups and deal effectively with situations where problems are poorly defined and require the "imaginative" utilization of available resources for their resolution.

The Nature of Man: the Seminar on Technology and Culture at M.I.T.

Massachusetts Institute of Technology, The Seminar on Technology and Culture, Program description available from Rev. John Crocker, Jr., Convener, 312 Memorial Drive, Cambridge, Massachusetts, 02139. The Seminar on Technology and Culture at M.I.T. is an independent and unofficial group of faculty and students gathered to stimulate study and discussion of the issues surrounding the interaction of science, technological values, and society. Seminar activites include weekly luncheon-discussions, and a series of year-long seminars called the "Images of Man." The seminar series deals with an understanding of the three types of human knowledge — the scientific, the personal, and the introspective; each form of knowledge has its own logic and realm of discourse. The program attempts to define the relationship of scientific knowledge to these other kinds of knowledge, since scientific knowledge has been the most "potent influence on human understanding in the last three centuries." This definition is important, the program's members feel, since "understanding the nature of man and his works has become a precondition for the survival of our species . . . . What we choose to believe about the nature of man affects our behavior, for it determines what each expects of the other." The seminar treats anthropological, biological, linguistic, psychological, literary, philosophical, and religious perspectives on mankind. The seminar also sponsors the Carl Taylor Compton lectures on the future of progress which focus on the process of industrialization in an attempt to describe how our society developed and now functions, and predict what the consequences of continued industrialization might be for America.

British Group to Monitor Human Consequences of Science.

Council for Science and Society, c/o Sir Michael Swann, Chancellor, Edinburgh University. Thirty-three eminent British scientists joined in July of 1973 to form a new Council for Science and Society. Chaired by zoologist Sir Michael Swann, the Council plans to examine specific areas of scientific inquiry, draw up a balance sheet of the social costs and benefits of these projects, and propose means to ameliorate the social costs. Initial Council projects will include examinations of research aimed at the control of human behavior and the prenatal determination of the sex of infants. The Council hopes to stimulate public debate through publishing reports of its findings. The Council is privately funded, and its members span the disciplines of medicine, physics, astronomy, psychiatry, literature, history, business, engineering, and the biosciences.

British Polytechnics.

Peter Scott, "Polytechnic Profiles," *The Times Higher Education Supplement,* 15 Oct. 71 on, a series of profiles of the 28 British polytechnics. The articles discuss the structures and histories of the various polytechnics, their academic strengths and weaknesses, their ongoing attempts at institutional innovation. Central to each profile is a discussion of the growing role of the humanities at the polytechnics. The school of arts and social sciences at several of the polytechnics are growing more rapidly than the institutes themselves. The changing image of the traditionally conservative polytechnic is the theme of the series of articles.

Engineering Education.
American Society for Engineering Education, monthly, October through May, Leslie B. Williams, Executive Editor, suite 400, One Dupont Circle, N.W., Washington, D.C. 20036. Annual nonmember subscription, $20.00.

Landmark Study of Engineering Education.
American Society for Engineering Education, "Liberal Learning for the Engineer," American Society for Engineering Education, 2100 Pennsylvania Avenue, N.W., Washington, D.C. 20037, (1968), 40 pp., $2.00. The Olmsted Report, as it is commonly known, differs from other studies of the state of engineering education in that it deals primarily with changing directions in the humanities and social sciences as they relate to the evolving character of society and the changing role of the engineer in society. It is the contention of the study that technology has forced modern man to make urgent decisions about his future. The engineer, seen as an independent and responsible policy-maker rather than as an employee of the rapidly expanding industrial system, plays a growing role in these decisions. Engineering education must be altered to equip the engineer for his new role within the system; it must emphasize the network of interrelationships that comprise the human environment. The study group found that while widespread revisions were occuring in the engineering curriculum of the country's major universities, the universities had not yet attempted any overall rethinking of engineering education, nor had they developed a coherent vision of what might be done to meet new needs. The report lamented the demise of general education on a university level as well as the segregation of liberal arts and engineering students and faculty.

Basic Engineering Education Requirements.
Goals Committee, American Society for Engineering Education, "Goals of Engineering Education," American Society for Engineering Education, 2100 Pennsylvania Avenue, N.W., Washington, D.C., Jan. 68, 74 pp., $2.00. "The rapid accumulation of new knowledge, . . . the accelerating pace of technological development, and the growing complexity of social, economic, and technical interrelationships in modern society demand a careful and continuing reappraisal of all educational practices." In this climate of concern, the Goals Committee of the American Society for Engineering Education in 1968 completed a comprehensive investigation of current trends in engineering education. The Committee found two major trends at work in engineering education; a search for unity of purpose and method in engineering education, and a trend towards broadening the content of technical studies. Engineering had assumed the role of a "liberal science", the report states, the counterpart in science and technology of

the traditional broadly-based education in the liberal arts. The growing demand for breadth in engineering education, however, tends to conflict with the increasing demand for skilled engineers. A partial solution to this conflict, the report suggests, might lie in raising the basic professional degree to the master's level, coupled with greater diversity of studies and extended off-campus and continuing studies. The report does not dwell on the need for increased humanistic studies in the engineering curriculum; however, it does observe that "engineering education must impart a thorough knowledge of the many non-technical aspects of modern life which interact significantly with the technical problems."

Workshop on Society and Technology.
Commission on Education, National Academy of Engineering, "Social Directions for Technology," National Academy of Engineering, 2101 Constitution Avenue, N.W., Washington, D.C. 20418. The report was produced by a workshop convened to develop strategies and mechanisms to redirect engineering education toward social needs. The report contains a concise summary of the problems and potentials of modern technology. The suggestions that follow are equally general. Meeting the challenge requires not only changes in engineering education but also new responsiveness on the part of governments to technological change, and strengthened relationships between industry and academia. The report calls for involvement of the public in making technological decisions by means of popular education for technological awareness. Decision-makers must have access to technological alternatives, and universities are particularly well-equipped to identify and make known such alternatives. Engineering capacity to incorporate social and behavioral considerations into design and analysis can be developed through broadened education. The report contains abstracts of case studies on transportation planning, housing, engineering education, and opposition to nuclear power.

A Manual for Problem-Centered Engineering Education.
American Society for Engineering Education, *Interdisciplinary Research Topics in Urban Engineering* (Washington, D.C.: The American Society for Engineering Education, 1969), 312 pp., paper, $5.00. This report is written primarily for university students "to assist in the development at universities of interdisciplinary systems engineering research in the delineated areas of urban engineering." (p. 4) The report concentrates on the three problem areas of urban transportation, urban housing, and the urban environment. The dimensions of each problem are discussed, as are current attempts at problem-solution and a description of Federal Government activities in the problem areas. Specific research topics are suggested and an extensive bibliography is included in each section.

Overview of Environmental Education.
Alan McGowan, "Education: University Programs," *Environment*, Vol. 15, no. 2, pp. 5, 41-2. Only fifty of the 2500 institutions of higher education in the U.S. have

environmental programs which are "significantly comprehensive and challenging," says the report. There are only a few schools which devote their entire programs to environmental education, such as the College of the Atlantic, in Bar Harbor, Maine, or the University of Wisconsin at Green Bay. Most schools merely add a program in environmental education to their liberal arts curriculum, and the program remains peripheral to the interested student's major field of study. Two barriers to the successful establishment of university programs in environmental education, the report states, are the difficulty of starting any interdisciplinary program, and the lack of support for all undergraduate research. Two successful environmental education programs are discussed at length: the Department of Environmental Studies at San Jose State College in California and the Program on Environmental Studies at the University of California at Santa Cruz.

College in the Image of Man.
Lloyd J. Averill, "Regaining the Colleges' Liberating Vision," *The Chronicle of Higher Education,* 2 Apr. 73, p. 12. The main problem besetting liberal arts colleges today is the result of the loss of a liberating vision in education. Institutional rigidity has resulted from the notion that liberal education is synonymous with some particular curriculum relationship or degree requirement. The diminishment of the value put on man in modern society is in part the result of this failure in liberal education. The essence of true liberal education must be a passion for man, a holistic style of teaching and learning, and an appreciation of human diversity. The author calls for a rebuilding of the college of human diversity, in the image of man, dedicated to human wholeness as an educational end and individual-oriented flexibility as educational means.

Humanistic Studies "Recivilize" America.
Robert F. Goheen, "America's Future and the Humanities," *Congressional Record,* 16 Dec '71, H12677-79. Originally appeared in *Foundation News,* Sept.-Oct. '71. Goheen stresses that humanistic studies are vital for America today in that they serve to redefine, to clarify, to reveal the transcendant importance of the "key words" which hold society together. A humanistic perspective defines these value-infused concepts, such as honor, in terms of human excellence, and this pursuit of human excellence must be a primary motivator in America's pursuit of technological progress. Goheen goes on to plead for a more imaginative, "relevant" approach to teaching the humanities and for greater foundation support of humanistic studies. He laments the disproportionate concern taken by foundations and government in scientific and technological education, and indicates that world events demand as great an awareness of the fine shades of right or wrong in American behavior as of the technical considerations involved in dropping a bomb. One feels, however, that although Mr. Goheen's concern for the revitalization and expansion of humanistic education is evident and that the connections that he suggests between humanistic studies and American self-awareness are indeed real, he has not figured out how to translate them into institutional action.

Federal Perspective on Change in Engineering Education.
Congressman John Brademas, Remarks before the American Society for Engineering Education, *Congressional Record,* 5 Dec '71, E13485-88. Congressman Brademas dwells on the pitfalls of technology as he sees them: the fallibility of technical systems, unwanted spillover effects, the lack of integration of goals among various fields of technology, and the lack of a system of priorities that keeps social considerations in mind. He believes these problems offer a challenge to our technological society, and believes engineering education to be the key to avoiding them. The goal of an engineering education must not be to make engineers into philosophers, he feels, but to provide "problem-focused education and research directed toward people — their needs and desire for a satisfying life in present surroundings." Such education can be pursued most effectively with federal help, and Brademas discusses the proposed National Institute of Education, the 1965 International Education Act, the National Foundation for the Arts and Humanities, and the Environmental Education Act, which he introduced into Congress, as the keystones of the federal effort. He further urges institutions of higher education to become more responsive to minority group needs and the directions of potential social change.

History and Structural Engineering.
David P. Billington, "Engineering Education and the Origins of Modern Structures," *Civil Engineering,* Vol. 39, no. 1 (Jan. 69). By studying the structures built during the late Nineteenth and early Twentieth Centuries, the engineering educator can integrate the humanities directly into his technical course on structural engineering. Billington believes that one way into the center of modern civilization is through its "tall, bare structures," for modern art and modern structure are intricately connected. Billington pursues his point with a series of illustrations. Not only did the development of steel encourage nineteenth-century engineers to create new technologies and new forms of structure, such as the Eiffel Tower, but steel appeared to the artists of that period as a symbol of their era. The outraged writings of Hugo and Merimee prompted a serious movement in France for the scientific study and preservation of its deteriorating medieval monuments. The Eads and Brooklyn Bridges, the visions of great engineers, stimulated the talents of front-rank American artists. In sum, Billington advises that if engineering education can be taught by reference to the masterworks of the recent past, it will result in the pursuit of that "humanistic ideal of involving the whole man in the quest for order and beauty."

An Engineer's Call for Working Relations with Humanists.
J. D. Horgan, "Technology and Human Values: The "Circle of Action," American Society of Mechanical Engineers reprint no. 72-WA/TS-4, 7 pp., 1972, $3.00. Reviews alternative formulations of the interactions of technology and society, explicitly rejecting von Neumann's view that technology is neutral in favor of a "circle of action" in which assessment of technology leads to modifications in operating values.

Studies of Culture and Technical Education.
Martin Green, *Science and the Shabby Curate of Poetry; Essays about the Two Cultures* (New York: W. W. Norton and Company, 1964), 159 pp., $5.00. A sensitive and thoughtful appraisal of the difficulties of including humanistic or liberal studies in technical education. Green says of students in British technical institutes, "The students', too, is a powerful, self-sufficient, somewhat embattled tradition. These students have the prejudice against literature (taken seriously) of all plain men in our society, plus the resentment against culture of those who have failed to get into a university (many did not get through grammar school), plus the pride of those who are trained in a rival, more masculine, more socially powerful discipline. It is no good approaching them with an offer of the truth, with the gospel of the true religion, as if they were pagans in some more primitive culture. . . . The only useful way to approach them is as fellow workers in another branch of a common intellectual tradition. But now many teachers of literature could do that?" (p. 115)

Poetry for the Student of Science.
T. R. Henn, *The Apple and the Spectroscope; Lectures on Poetry Designed (in the Main) for Science Students* (New York: W.W. Norton, 1966), 166 pp., paper, $1.95. The lectures which comprise this volume were delivered over twenty years ago to science students at Cambridge; they attempt to pave the way to an interest in literature through an understanding of the function and technique of poetry. The lectures are analytical and critical, and the author is concerned to bridge the gap between the sciences and the humanities, by pointing out to his students that they do not have to modify any of their mental processes in approaching literature. All great art, he argues, is "ultimately moral, as having a positive and constructive relationship to the sum of our attitudes toward life" (p. 137) and may profitably be included in technical education.

Higher Education and Values in Context.
Maxwell H. Goldberg, *Design in Liberal Learning* (San Francisco: Jossey-Bass, 1971), 188 pp. $8.75. A discerning analysis of the tension between tradition and sympathy for change manifest throughout liberal arts education, continued with a specification of needs to include symbolic value communication processes in building "competence in a technotronic age" which "calls for understanding the dislocations that disturb wholeness and are created deep in man's psyche by technological feats such as those which caused the holocaust at Hiroshima."

Technology as an Institution: Social and Cultural Aspects

Belinda Barrington
and
Philip C. Ritterbush

Technology at the Value Nexus: Charles Frankel.

Discussions of technology in contemporary society have tended to cluster at two opposing poles of policy. There are some who maintain that "more" technology will overcome the deficiencies of previous applications and others who regard technology as the incubus of a doom from which man can never rid himself. Both points of view regard the character of technology as fixed and differ only as to whether they prescribe more or less. In neither view, as usually expressed, is technology a social process, being considered instead as machines and industrial processes. If one takes the wider view, that technology consists of processes for achieving results, it would include management techniques, the practical arts, and methods of analysis and operations. Less abstractly, technology is what engineers practice, and that is by no means restricted to mechanical processes. It includes planning, analysis, and design, not just of mechanical systems, but of human organization, renewable resources, agriculture, and even language.

Once matters are viewed this way, it becomes meaningless to speak of less technology or more. Technology is instead a pervasive element of social organization and one of our major contemporary institutions. As such it is deeply implicated in questions of social purpose and politics. That this is so was a leading contention of an important interpretation of social thought, *The Case for Modern Man* (N.Y.: Harper & Brothers, 1956), by Charles Frankel. That book ascribes the central involvement of technology in our society to its dominant political outlook — "liberalism." Some of the distinguishing attributes of this outlook are faith in reason, "a general predisposition in favor of reform" and the solution of social problems, insistence that the state observe moral standards, and, above all, the idea "that science can be the central organizing agency for modern society." (p. 34) "Liberalism was the social movement that spoke most distinctively for modern man's sense of the new powers which technology and science had brought him." (p. 29)

It is sometimes said that technology embodies no values, which is to ignore the social context that has had so much to do with the directions of its development. Frankel keeps this in view throughout his book, whose principal purpose is to reassert the validity of traditional liberalism in the face of four challenges, which he locates in the thought of Maritain, Niebuhr, Mannheim, and Toynbee. Thus the book can serve as a general review on technology and social thought, and it might be pointed out that few more recent books have gone beyond it as a summary of the questions presented by technology.

Maritain held that society could not reflect a moral order without absolute values, to which Frankel counters that institutions should derive their authority from their tested capacity to serve living human interests rather than on claims of special access to external truths. Thus it is futile in Frankel's view to look for constant values as determinants of the technological enterprise.

Niebuhr considered that evil was an inevitable aspect of man's capacity to act, which would make it an inescapable attribute of technology. Arguments for a moratorium on technology imply that evil consequences are inevitable. Frankel counters by asserting that man can progress incrementally by solving social problems and that no definite limit can be drawn beyond which he cannot make his way.

Mannheim stressed a social conditioning of thought that seemed to make objectivity unattainable. Frankel replies that objectivity is grounded in the processes of mutual criticism that govern science, and that is the task of theory to impel intellectual advances that can later be verified. Technology incorporates continuous criticism as the driving force toward innovation, and thus escapes Mannheim's strictures on the narrowly conceived rationality of the Academy.

Toynbee held that technology, unlimited use of science, and innovations in law and social planning tended to lead to a universal state that would inevitably break down for lack of the transcendant values that alone can keep a civilization alive. Frankel doubts that man must adhere to original values and submits instead that changes continuously take place — that the "partial, piecemeal reform of social institutions" can bring about meaningful change.

Frankel considers technology the dominant change force in the modern world, one of the "tidal movements in our institutions . . . the fundamental dynamic element in modern society."

The decision as to when, where, and how to introduce a technological change is a *social* decision, affecting an extraordinary variety of values. And yet these decisions are made in something every close to a social vacuum. Technological innovations are regularly introduced for the sake of technological convenience, and without established mechanisms for appraising or controlling, or even cushioning their consequences." (p. 198)

The institutions that deal with values are fragmented so that decisions are taken in a partial and contradictory manner, for want of a multipurpose institution where a wide range of values could be dealt with. "The problem cannot be met by reminding engineers of their social responsibilities or by calling conferences to discuss the human use of human inventions. The problem is institutional." (p. 199) In Frankel's view the main need is to make the social policy process more responsible. This will necessitate a systematic effort to relate individuals to social authority and the development of a new theory of society. Thus the problem posed by technology is not an engineering problem but a broad-based complex of social difficulties. Even so, technology remains the supreme example of man's determination to employ the full range of his powers:

As its crowning symbol [the revolution of modernity] developed a radically new outlook on human destiny, which saw the meaning of history in terms of the progress of the human mind, and held that human history could be made to follow the direction that men chose to give it. Prometheus was the first modern. The revolution of modernity proposed to put men squarely on Prometheus' side. It is a unique venture in human affairs, and one can only relieve the strains and tensions it has created by taking it seriously. Our disappointments are real. But they are real because our powers are great and our expectations legitimately high." (p. 209)

Technology Interpreted as a Social Institution.
John G. Burke, ed., *The New Technology and Human Values* (Belmont, California: Wadsworth Publishing Company, 1966), 408 pp., paper, $3.95. By "new technology" Burke means the interaction between basic science and applied science; the book is concerned with a second convergence — the impact, past, present, and future, of technology on society. The book begins with an expository, general section on the impact of science and technology on society, including a brief historical perspective and a longer section on resulting changes in educational methodology and requirements. A second group of essays considers the nature of automation and resulting problems of employment and leisure. Additional essays deal with questions of population and individual freedoms and the interaction of technology with the policy-making process.

Technology in the Social System.
Aaron W. Warner, Dean Morse, and Thomas E. Cooney, eds., *The Environment of Change* (Columbia University Press, 1969), 186 pp., $8.50. The lectures presented here address themselves to the general subject of social change in modern society and the place of technology in the context of societal change. The authors speak of the scientific roots of technological change and of recent developments in biological technology. They describe changes wrought by modern technology on employment patterns, on personal and social values, and on behavior, and of the role of technology in destroying human diversity and in imposing American patterns on other nations in the developing world.

160

Science, Organization, and Social Policy.
Quintin McGarel Hogg, *Science and Politics* (Chicago: Encyclopaedia Britannica Press, 1964), 110 pp., $2.95. "Government is a financing, coordinating, participating, training function. Over a large field [like science] it is not, and cannot be, directory and executive. These are functions in which the main role ought to be played by industrialists, educationalists, teachers, and scientists themselves." (p. 19) The author, then British Minister for Science, discourses on the organization of science, the educational background necessary for a modern scientist, the opportunities afforded by an active industrial sector for the encouragement of science, and the challenge presented by international communism to spur scientific development.

Technology and Government.
C. P. Snow, *Public Affairs* (New York: Charles Scribner's Sons, 1971), 224 pp., $6.95. The subject, although not the thesis, of Lord Snow's widely noted "Two Cultures" lectures of 1959 (here reprinted), has presented ever larger claims on attention. The thesis, that those educated in a literary culture could not understand technology, has fortunately been shown to be untrue. They have understood its social and environmental effects, which are part of technology as an institution, better than engineers and civil servants, for whom it has usually been only a technique. Public policy will be powerless to correct the situation until it deals with technology as a social institution, which perhaps accounts for Lord Snow's failure to offer specific suggestions for public action and the rather helpless pessimism he professes in this collection of lectures and essays.

Technology as Its Own Antidote.
Werner von Braun, "Hostility to Technology Is Irrational," *Congressional Record*, 14 Oct. '71, E10845-46. Von Braun charges culturally oriented critics such as Lewis Mumford or Charles Reich with presenting a lopsided view of science and technology. Such sentiment frequently takes as its premise a deep concern with conservation, a clean environment, or improving the quality of life. Yet the solutions to these same social problems depend in part upon our technological capabilities to solve them. Not only are the mechanisms developed by science and technology thus valuable to social welfare, he argues, but also the creative potential of science and technology broadens man's horizons.

Technology Interpreted as a Benign Social Influence.
Amitai Etzioni. "Humane Technology," *Science*, Vol. 179 (9 Mar. '73), p. 1. While certain technologies do promote an "impersonal, efficiency-minded, mass-production" society, Etzioni argues that other technological developments are essential for a more humane society. Some technologies eliminate drudgery; other technologies contribute to the solution of social problems, like instructional television and seat belts. Etzioni cautions us against rejecting all technologies; the world of the future will not be an idealized return to nature but rather a more

humane world reached through technologies geared to advance our true values. Thus pollution will be reduced through the development of less polluting technologies, leisure time will be augmented as more tasks are performed by machines supervised by other machines, and communication and transportation technologies will enable an ever-increasing number of people to enjoy the benefits of civilization.

Technology Review.
Monthly, John I. Mattill, Editor, Massachusetts Institute of Technology Room E19-430, Cambridge, Massachusetts 02139. Annual subscription, $9.00 domestic, $10.00 foreign.

Science, Technology, and Society.
UNESCO, *Impact of Science on Society,* quarterly, Jacques Richardson, Editor, annual subscription, $4.00, from UNESCO, 7 Place de Fontenoy, 75700 Paris, France or its U.S. publication distribution center, P.O . Box 433, New York, N.Y. 10016.

Technology Assessment: A Pioneering Institutional Review.
Vary T. Coates, *Technology and Public Policy; The Process of Technology Assessment in the Federal Government,* 3 vols., Program of Policy Studies in Science and Technology, George Washington University, Washington, D.C. 20006. Reviews the activities of 86 offices in the Federal Government. Housing, biomedicine, space, transportation, and mineral resource extraction are cited as subject areas deficient in both the quality and scope of effort. In recommending that federal agencies serve as frameworks for the development of technology assessment the report tends to overlook their institutional bias in favor of their functions as they define them, usually more narrowly than the social function they serve, federal programs in transportation rather than transportation as a whole, for example.

A New Social Dynamic in Technology.
Marvin J. Cetron and Bodo Bartocha, eds., *Technology Assessment in a Dynamic Environment* (New York: Gordon and Breach, 1973), 1036 pp., $45.00. This collection of case studies and analytic essays presents a comprehensive overview of technology assessment and forecasting written by experts on technology assessment around the globe. The book begins with an overview of technology assessment in several countries and continues with a survey of analytic methods available for technology assessment, including cost-benefit analysis, use of societal indicators, and problems of measurement. This survey is supplemented by case studies, including consideration of questions of policy planning and resource utilization. It concludes with conjectures regarding the internal and external influences on technology assessment as they relate to its possible future development.

International Society for Technology Assessment.
10 Churchillplain, P. O. Box 9058, The Hague, Netherlands. A new world membership organization, annual subscription $20.00 ($10.00 for students), which entitles members to receive *Technology Assessment* (issued quarterly). U.S. information office, Suite 510, 1015 18th Street, N.W., Washington, D.C. 20036. An international congress on technology assessment was convened under its auspices in 1973. Journal subscriptions also available from the publisher, Gordon and Breach, for $16.00 to individuals and $53.00 to institutions.

Technology and Public Policy.
National Science Board, *The Role of Engineers and Scientists in a National Policy for Technology* (National Science Foundation, 1972), xi + 48 pp. $0.45.

Technology Assessment and Public Policy.
Raphael G. Kasper, ed., "Praeger Special Studies in U.S. Economic and Social Development," *Technology Assessment; Understanding the Social Consequences of Technological Applications* (New York: Praeger, 1972), 291 pp., $16.50. The volume contains ten papers presented at a series of seminars on the process and mechanisms of technology assessment conducted by the Program on Policy Studies in Science and Technology at George Washington University. Paper topics reflect the program's concern with the role of technology assessment in Federal Government policy planning activities, and therefore deal with the relationship of technology assessment to the Congress, the Executive, the Food and Drug Administration, and the federal transportation authority. Additional papers assess the state of technology assessment methodology and management and discuss the role of technology assessment in widening citizen control over the directions taken by technology.

Case Studies of Technology Assessment.
Martin V. Jones, *A Comparative, State-of-the-Art Review of Selected U.S. Technology Assessment Studies* (May, 1973). The MITRE Corporation, Westgate Research Park, McLean, Virginia 22101. 92 pp. Analysis of the kinds of institutions participating in technology assessment and the methods employed.

Architecture and Social Change.
Gerald McCue, William Ewald, *et al., Creating the Human Environment,* A report for the American Institute of Architects (University of Illinois Press, 1970), 339 pp., paper, $4.95. "Technology is ... the vision of the future most easily communicated and most easily marketed. . . We have no comprehensive social-economic-physical conception of the total future human environment, or of how to build it beginning now and phasing through the transitional period of the next two to fifteen years into the technological future." (p. 8) Man has time to plan his future environment before inevitable technological trends overtake him, the report argues, but can he learn the foresight necessary to do so? The report is con-

cerned with the recent past and distant future of the profession of architecture as the "art and science which bridges both humanistic and scientific values" (p. 311) in planning for man's future. Concepts of change within the building industry are surveyed for the purpose of forecasting the shape of the man-made environment within the next fifteen years.

How Can Engineering Improve the Man-Made Environment?

American Society for Engineering Education, "Industry-Engineering Series 1-4," *Our Technological Environment; Challenge and Opportunity*, proceedings of 23rd Annual College-Industry Conference, American Society for Engineering Education, Washington, D.C. Jan. 1971, 224 pp., paper, $2.00. Leaders of education and industry were asked to present their perceptions of man in a technological world. How can man alleviate the negative environmental consequences of resource utilization? How can he deal with the problems of urban concentration? How can he develop transportation and information systems better able to guide the future development of society? The comments that comprise this conference summary are general and far-ranging, dealing with the quality of life rather than the specifics of technology in operation.

The Futurist.

Bimonthly, Edward Cornish, Editor, published by the World Future Society: An Association for the Study of Alternative Futures, Post Office Box 30369 Bethesda Branch, Washington, D.C. 20014. Annual subscription $10.00. Special institutional memberships, $100.00 per year.

The Future as a Social Invention.

Dennis Gabor, *The Mature Society* (New York: Praeger, 1972), 208 pp., $7.50. The time has come, the author argues, for man to put his creative energies into "social invention" to improve the quality of life. In this sequel to *Inventing the Future*, he sketches his version of a mature society, "a peaceful world on a high level of material civilization, which has given up growth in numbers and in material consumption but not growth in the quality of life," (pp. 3-4) while maintaining the maximum amount of freedom compatible with social stability. Man is not required to plan now the details of future civilization, but he should, the author argues, anticipate dangers that an Age of Leisure will bring. In an inventive and scholarly treatise, Gabor diagnoses the ills of our society and suggests a variety of far-reaching changes in education, employment practices, fiscal policies, and attitude that will instill a sense of personal responsibility and individual discipline and a drive toward excellence instead of quantitative growth in the economy. Sweden serves as the author's model for the improved world of the near future: a hopeful but not uncritically visionary prospect.

The Components of Technology and Technological Leadership.

C. S. Draper, "Technology, Engineering, Science, and Modern Education,"

164

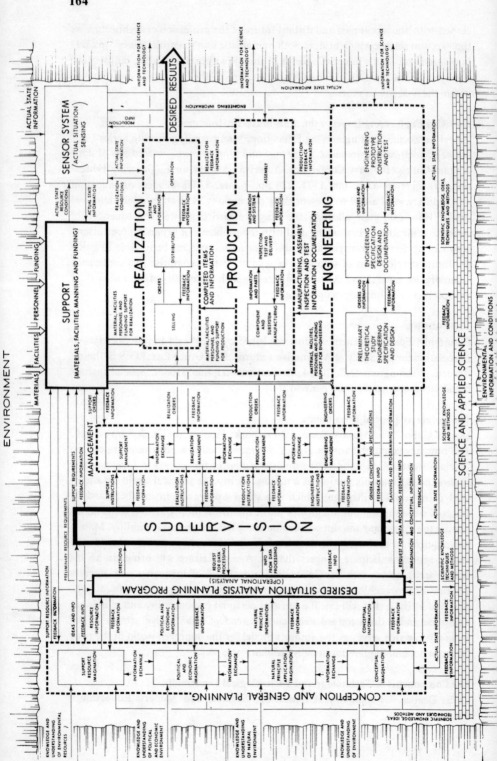

Leonardo, Vol. 2 (Pergamon Press, 1969), pp. 147-153. The ever-increasing complexity of technology gives rise to needs for effective social leadership. Problems can no longer be handled satisfactorily by a pooling of individual talents, since fragmentation of discipline and interest make cooperation and effective communication difficult for comprehensive multidisciplinary projects. The author feels that a new profession is needed for leadership in technology which combines a traditional technical capacity for systematic problem-solving with an orientation toward perceiving the broad ecological context of technological activity. The author elaborates on the personal qualifications necessary for such leadership and proposes a new form of education designed to develop these qualities. Such an education would continue instruction in science, mathematics, and the humanities, but would emphasize elements of technology and engineering that distinguish these disciplines from science and applied science by orienting them toward responsible action. The institutions developed to provide such instruction should devote a significant amount of time to preparation for real-world activity, as do institutions for legal and medical education. The author cites the Instrumentation Laboratory at M.I.T., which he directed for thirty years, as a useful model.

Professional Values and Social Accountability in Technology.

Committee for Social Responsibility in Engineering, *Spark,* semi-annual, with an annual subscription of $10.00. 475 Riverside Drive, New York, N.Y. 10027.

Military Aspects of Technology as an Institution.

Adam Yarmolinsky, A Twentieth Century Fund Study, *The Military Establishment; Its Impacts on American Society* (N.Y.: Harper & Row, 1971), xiv + 434 pp., Colophon Paperbound, $3.95. This judicious and exceptionally wide-ranging study of "the largest institutional complex within the United States Government" is concerned primarily with its exaggerated influence on public policy, which Yarmolinsky considers to be a consequence of insufficient public understanding of the civil functions of government, resulting in a resort to using the U.S. military as a global police force and budgeting for necessary national expenditures, as for the support of research, through inappropriate military agencies. The bias toward glamor technology is attributed to the popularity of hardware procurement contracts within the Congress. The book demonstrates the influence that the institutional setting exerts over the social character of technology in its most lavishly funded and strongly emphasized areas of development.

Institutional Bias in Social Science.

Irving Louis Horowitz, ed., "Transaction Studies in Social Policy," *The Use and Abuse of Social Science: Behavioral Science and National Policy Making* (New Brunswick: Transaction Books, 1971), 350 pp., $8.95. A major study of the political uses of social science research, this book suggests that social science today is often used to legitimate government decisions made on political grounds.

166

A summary of papers presented at a conference on the relationship between social science and public policy held at Rutgers in 1969, the book addresses four major issues: the autonomy of social science versus its utility, relevance, and the commitment of the scientist to do policy-relevant work; the differential availability of research results to opposing parties in the formulation of public policy; the relationship between the design and analysis of social systems; and the lack of readiness of traditional social science to perform policy-relevant work.

Government Encouragement of American Science in the Pre-Industrial Era.
Forest G. Hill, "Formative Relations of American Enterprise, Government and Science," *Political Science Quarterly*, Vol. 75 (1960), pp. 400-419. Early governmental promotion of American science and private enterprise was largely a response to the practical needs of a pre-industrial era. Public action for the benefit of transportation, industry, agriculture, and national defense both required and stimulated scientific advances and thus encouraged the development of private enterprise and scientific educational institutions. The technical bureaus of the Navy and the War Department were the instruments for promoting several kinds of investigative and technical endeavors that were not related to national defense, including the importation of technical knowledge from Europe. The Smithsonian, in its early years, worked in close liaison with such governmental agencies as the Coast Survey and Engineering Department, as well as such private scientific groups as the AAAS. After the Civil War, private institutions began to take over the government lead in civil engineering, geology, minerology, astronomy, meteorology, agronomy, and ethnography, and the newer disciplines of chemistry, physics, and industrial technology. The author finds a closer parallel between today and the pre-Civil War, rather than post-Civil War, era in terms of the close supportive relationship between government and scientific enterprise.

The Development of Functional-Performance Institutions.
Michael Michaelis, "Humane Technology for Business Betterment," *The Conference Board Record*, Vol. 10, no. 2 (Feb. 73), pp. 56-60. Man must create new institutional relationships to spur innovation for social betterment, Michaelis feels, with the same energy that he has previously put into creating new knowledge. New programs involving the National Science Foundation and the Department of Commerce to test a range of incentives designed to encourage technological innovation in the private sector are models of the institutional arrangements that he advocates. Equally important, Michaelis says, is the development of a new mode of perception by business management that he calls a "functional performance" orientation. Industry must be reorganized in order to sell "performance" to meet the functional needs of society rather than "products" he explains. The consumer would then elect to purchase a number of units of "performance" rather than the separate component products and partial services now offered by various institutions to meet social needs. This holistic approach to social services, he believes, would encourage continuous technological

and institutional innovation, which would include the fostering of business practices that are capable of higher risk-taking, and of responding in anticipation to need rather than in reaction to crisis. The resulting technologies would be humane, in that they would be significantly influenced by considerations of societal good, presumably, and they would be evaluated in terms of "systems performance," which would tend to optimize allocation of resources. Michaelis cites the Health Maintenance Organization as a possible model for a functional-performance institution.

Research Policy and Social Values.
Dr. Hertha Firnberg, "The Social Function of Science Policy," *Science Policy*, Vol. 2, no. 1 (Jan.-Feb. 73), pp. 5-6. Science policy should be regarded as an integral part of national and international social, economic, and cultural policies, Dr. Firnberg argues. Science policy provides a framework that helps identify other policy objectives by providing a model of the future which protects human values and offers alternative paths to reach these objectives. Research policy bears the responsibility for providing the funds necessary for research and development activities, and also for ensuring that the problems and objectives of social, economic, and science policies are publicly discussed and that funds are spent in accordance with the relative importance of these objectives. In the spring of 1972 the Austrian government published the *Austrian Guidelines for Research*, prepared by the Federal Ministry of Science and Research. The *Guidelines* addresses the principles underlying Austrian research policy, fits research and development into the hierarchy of social values, and contains medium and long-range plans for research policy. Its implementation will call for close cooperation among government, industry, the scientific establishment, and an informed and critical public, but does not call for the creation of new institutions to facilitate such cooperation.

FASST: Federation of Americans Supporting Science and Technology.
Robert J. Morrissey, "Technology's Youthful Activist," *Congressional Record*, 26 July 72, E7101-02, reprinted from *Bee-Hive*, the magazine of United Aircraft. FASST is an organization centered at the University of Michigan, created to counter the growing anti-technology sentiment of students and society. David Fradin, an undergraduate engineering student, founded the organization in 1970 to combat the sentiments that brought about the defeat of the SST. The group has continued to provide a forum for the discussion of technological problems and the exchange of information among individuals interested in technology. Fradin believes that anti-technology sentiment can be ameliorated by distributing balanced information on the goals and activities of technology and that universities are particularly well equipped to perform this function. He further feels that aerospace institutions, in particular, should concern themselves more with the dissemination of accurate technological information to universities and other public forums.

Widening the Social Base of Technology.
Congressman J. Edward Roush, "Consumers of Technology," *Congressional Record,* 18 Nov. 71, E12376-78. How can we justify vast governmental expenditures in research and development in the face of gaping social wounds, immediate environmental concerns, and transportation and housing crises, Roush asks. Who will be the consumers of the technologies developed with public funding? Roush addresses these issues from the standpoint of a former member of the House Science and Astronautics Committee, witth a strong belief in the desirability of technology transfer. Roush finds that Federal Government agencies have generally failed to widen the use of technology . In an effort to improve their performance, he has introduced legislation to establish an Office for Federal Technology Transfer to coordinate, reuse and constructively adapt knowledge and technology coming from federal expenditures for defense, atomic energy, and space research.

Federal Support and Direction for Technology Utilization.
Senator Peter Dominick, "Remarks on Civilian Science and Technology Policy." *Congressional Record,* 10 May 73, S8754-56. Senator Dominick speaks of the importance of federal support for science and technology, but he is concerned about the lack of federal programs to apply science and technology to civilian needs. He proposes several new institutional arrangements for the planning, analysis, and development of programs and policies aimed at the solution of civilian problems, such as health care, public safety, and education. He suggests that the National Science Foundation both conduct and support studies designed to assess national problems and alternative solutions, and compile information on civilian research and development activities, paying particular attention to the role of the Federal Government in influencing the "economic, sociological, legal, administrative, and institutional factors" which affect the application of science and technology to civilian needs. He further suggests the establishment of an Intergovernmental Science and Technology Advisory Council within the Foundation to direct these efforts and to provide technical assistance and funds to state governments engaged in local efforts to apply technology to civilian needs.

Taming Technology.
J. Herbert Hollomon,, "Technology in the United States: The Options Before Us," *Science Policy Reviews,* Vol. 5, no. 3 (1972), pp. 2-13. This article is based on two more extensive papers. It summarizes social problems and discusses the policy alternatives that man might choose for the social control of technology. "Taming" technology would result, Hollomon says, in continuing improvement of the quality of human life. To this end, Hollomon advises, technology and science policy must incorporate a growing appreciation of humane values and new norms of collective action. Hollomon enumerates eight policy options that are available in planning new directions of technological development: one can allow present trends to

continue, directly or indirectly support private technical efforts, improve the service sector of the economy by innovation and diffusion, support the training and relocation of displaced workers, financially support high risk ventures, improve the process of technology transfer, and quantify and ameliorate the social consequences of technological change. The article discusses the positive and negative consequences of supporting each policy option, and makes specific practical suggestions about the mechanisms needed to pursue each. Hollomon stresses that the options that we may choose are contingent upon our acquiring a deeper knowledge of the functioning of our society, in order better to understand the consequences of our present patterns of technological growth.

History of the Role of Mechanization in Life.
Siegfried Giedion, *Mechanization Takes Command; a Contribution to Anonymous History* (New York: W.W. Norton, Inc., 1969, first published Oxford, 1948), 743 pp., $4.95. In this pioneering historical account of mechanization, Giedion (at one time associated with Gropius in the Bauhaus) presents a summary account of its development from handicraft to the assembly line and describes its penetration of agriculture, the food industry, and the home. His aim was to trace limits beyond which human values must be respected: "The coming period must bring order to our minds, our production, our feeling, our economic and social development. It has to bridge the gap that, since the onset of mechanization, has split our modes of thinking from our modes of feeling." (p. v)

Technology and Consumerism.
Daniel Boorstin, *The Americans; the Democratic Experience* (New York: Random House, 1973), 717 pp., $10.00. The productivity drive of the American economy has made of technology a daemonic flow of ever more goods, overwhelming consumers with possessions and distracting technical capacities from conditions that might otherwise be remedied. So has arisen what Daniel Boorstin calls, in an important new interpretation of the history of American consumer goods, "the new unfreedom of omnipotence."

Allocating the Dividends of Productivity.
Fred Best, ed., "The Human Future Series," *The Future of Work* (Englewood Cliffs, New Jersey: Prentice-Hall, 1973), 179 pp., $5.95. This anthology documents the current changing concept of achievement and goal-expectations of the American worker. The modern worker increasingly looks past the acquisition of material wealth toward "self-actualization," "social-belongingness," and self-esteem. The evolution of the structure of work in America will be determined by the change in priorities of such human needs. The essays in the book outline the evolutionary history of work, identify human felt needs, speculate on specific conditions and goals of work in the future, and present four alternative conditions that work might assume in the future.

The Problem of Leisure: A Cultural Impact of Technology.
Max Kaplan and Phillip Bosserman, *Technology Human Values and Leisure* (N.Y.: Abingdon Press, 1971), 256 pp. Contributions to a conference sponsored by the Center for Studies of Leisure at the University of South Florida, Tampa, mostly viewing leisure phenomena imaginatively in a cultural context. The productivity gains achieved by technology create free time; leisure is its culturally meaningful use.

Work Satisfaction: an Index to Technology Impacts.
U.S. Department of Health, Education, and Welfare, *Work in America* (M.I.T. Press, 1973), 262 pp., paper, $2.95. An H.E.W. task force reports on work in the lives of American adults. Sources of the current large-scale blue and white-collar worker dissatisfaction are discussed, as are its consequences in physical and mental health complications and decreased productivity. The report concludes with suggestions for the alleviation of the problem, including the redesigning of jobs, worker retraining, and change in Federal Government work strategies.

Career Orientation and Social Values.
Sanborn C. Brown, ed., *Changing Careers in Science and Enginnering* (M.I.T. Press, 1972), 349 pp., $10.00. The book is directed toward engineers and scientists who feel their careers threatened by changes in national policy or who merely desire a change in career orientation. The engineers, educators, and economists who delivered the lectures upon which the book was based forecast the manpower needs in the various technical professions over the coming decade, discuss the problems of personal adjustment to a new career, and relate changing patterns of engineering employment to national politics.

The Transformation of Work and Learning.
William W. Brickman and Stanley Lehrer, eds., *Automation, Education, and Human Values* (New York: Thomas Y. Crowell Company, 1969), 415 pp., paper, $2.95. "What are the humanistic implications for education of the pervasive impact of exponentially accelerating technological change?" (p. 12) In a world where change is confusingly rapid, education must be redesigned to keep pace with human needs. This book presents arguments for a raising of the broad educational level of the èntire work force. It speaks of the benefits of education for cultural enrichment, of an education preparing men to work in a fluid and dynamic environment. Furthermore, man should change the vocabulary and philosophy of work, the authors suggest, considering as work "any effort leading to the development of potentiality." (p. 346)

The Need for Institutional Transformation of Technology.
Eugene S. Schwartz, *Overskill; the Decline of Technology in Modern Civilization* (Chicago: Quadrangle Books, 1971), xi + 338 pp., $8.95. The author, former

senior scientist at the Illinois Institute of Technology, challenges the belief that more technology can devise remedies for the side-effects of previous technology.

Some Recent Views of Technological Man.
William Kuhns, *The Post-Industrial Prophets; Interpretations of Technology* (New York: Weybright and Talley, 1971), 280 pp., $6.95. The author distills and interprets the thought of Mumford, Giedion, McLuhan, Fuller, Wiener, Ellul, and Innis. The final section of the book explores the hopeful posture of social engineers and systems designers on the potential of technology for developing a better world.

Call for a Movement to Apply National Conscience to Redirect Technology.
Erich Fromm, "World Perspectives," *The Revolution of Hope; toward a Humanized Technology* (New York: Bantam, 1968), 178 pp., paper, $0.95. Fromm's interpretation hinges on his concept of "the system man" — a synthetic merging of human character structure, qualities and potentials which he believes may overcome contemporary social problems. Fromm's discussion of the human situation leads him to consider the potential for a better future reached through a humanized technology, which would entail a humanization of planning, grass-roots activation of human energies, a humanization of consumption patterns, and psycho-spiritual renewal.

Legal Limits on Technology.
David Loth and Morris L. Ernst, *The Taming of Technology* (New York: Simon and Schuster, 1972), 256 pp., $6.95. The authors provide a sketch of the evolution of legal controls over developing technologies in a number of areas, including consumer safety, space law, and meteorological manipulation. By means of a series of case studies, they discuss both international and domestic statutes which have been used in the past to remedy threats to the quality of life as diverse as errors in the personnel files of computer banks and dangerously lax safety limits for radiation. The authors argue that man will be exploited by the means he has for his own progress unless technology is brought under strict legal supervision.

Mechanisms for the Social Control of Technology.
Michael S. Baram, "Social Control of Science and Technology," *Science,* Vol. 172 (7 May 71), pp. 535-39. Baram argues that we should develop preventive and *a priori* controls on the direction and utilization of science and technology, rather than continue to put our efforts into remedial solutions for technologically-induced problems. Our present system of controls, Baram says, has failed; the legal system has proven particularly unresponsive to new social conditions, and both legal and private controls — crusaders and citizens' groups — operate retro-actively, and interfere only in the advanced stages of production and application of technology, after massive commitments of resources have already been made. The governmental agencies assigned control over technology are beset, Baram

172

states, with fragmentation of aim and process, conflicting loyalties, and bureaucratic inertia. Nor does the author place much confidence in the potential of scientific peer groups to function as humane social controls, entrenched as they might be in narrow disciplinary concerns. Baram places hope in legislation and litigation. All agencies should be subject to legislative controls modeled after the National Environmental Policy Act, which requires that "unquantified environmental amenities and values . . . be given appropriate consideration in decision-making along with economic and technical considerations." Further legislation should be directed towards the regulation of federal procurement and contractor activities. Public financial support should be given to citizens' groups engaged in activities in the public interest.

Conceptual Model for Technology Assessment.
Michael S. Baram, "Technology Assessment and Social Control," *Science*, Vol. 180 (4 May '73), pp. 465-473. "Established and ineffective patterns of post-hoc legislation, regulation, and litigation" are in part the cause, Baram says, of current problems of consumer protection, occupational health and safety, and the quality of urban life and the environment. Baram outlines here the necessary considerations for the development of new coherent frameworks for technological planning and decision-making. Technology depends on developments that occur in four interrelated processes, he notes: basic research, applied research, the development of prototypes for testing and experimentation, and production and utilization. Large economic and social commitments are made during the development and experimentation phase, which lends a measure of inevitability to the technological advance. Baram catalogues the resources for and effects of technological advance and the influence of decision-makers on both of these factors. Baram stresses that decision-makers are influenced not only by their perceptions of feasibility and probable effects and their access to information, but also by citizen response to program effects. The model Baram proposes is useful in that it articulates a system for the rational development of technological policy. In conclusion, Baram discusses certain reforms attempted recently, paying particular attention to the experiences of the Environmental Protection Agency and to procedures to enhance the flow of balanced information on technological developments to the public.

The Social Environment and Its Effects on Technology.
J. G. Crowther, *Science in Modern Society* (New York: Schocken Books, 1968), 403 pp., $8.00. A comparison of British and U.S. science policy in the context of social thought by a distinguished senior commentator. He champions "social change which will release new forces to inspire scientists and technologists." (p. xvii)

Who Controls Technology?
Anthony Wedgwood Benn, Myron Tribus, Daniel C. Drucker, and Alan G.

Mencher, "On the Control of Science: Four Views," *Science and Public Affairs: Bulletin of the Atomic Scientists,* Vol. 27, no. 10 (Dec. 71), pp. 23-38. These four essays address various aspects of the control of technology and the direction of the techno-scientific revolution. Topics include the impact of technology on governmental structures, the need for a new social invention to secure the consent of those affected by technology, the optimal role of the engineer in the establishment and the social deployment of science. The papers were originally presented at a Northwestern University Colloquium on "The Control of Science for Civil Needs" on April 12 to 13, 1971.

Law, Value Implementation, and the Control of Technology.
Laurence H. Tribe, "Legal Frameworks for the Assessment and Control of Technology," *Minerva,* Vol. 9 (1971), pp. 243-255. An interpretation of the role law may play in giving effect to the recommendations of technology assessment.

Cautions on Technology.
Richard D. Lamm, Colorado State Representative, "Technology Assessment and the Brave New World of the Future," *Congressional Record,* 17 Dec. 71, E13756-58. The accelerating rate and magnitude of technological change, Lamm cautions, has made it necessary to build a new system of checks and balances on technological development to ensure that every possible and identifiable consequence of technology is taken into account. The new system must emphasize the long-term consequences of technological change; Lamm mentions the Office of Technology Assessment as a model for this aspect of the new system. The proposed system would require the development of a better equation to include uncertainty in the decision-making process; he does not offer suggestions as to how this equation might be developed. A third improvement Lamm proposes would be stronger and more effective legal controls, which take into account the possibility of future harm and do not merely require proof of harm done. Lamm dwells on the failures of decision-making in the Project Plowshare Program of the Atomic Energy Commission, criticizing the Agency's stress on the short-term benefits of nuclear power to the slighting of its risks, and he notes the possible inconsistency of assigning to one agency both promotion and control of a technology.

The Concerned Engineer: an Introduction.
L. Daniel Metz and Richard E. Klein, *Man and the Technological Society* (Englewood Cliffs: Prentice-Hall, 1973), 180 pp., paper, $5.00. Written to introduce students to the world of engineering, this book offers both a subjective interpretation of the role of the engineer in society and a comprehensible summary of several techniques of modern "interdisciplinary" engineering, including optimization, modeling, cybernetics, and feedback.

Technology and the Use of Power.
William G. Carleton, *Technology and Humanism; Some Exploratory Essays for*

174

our Times (Vanderbilt University Press, 1970), 300 pp., $12.50. Technology is recognized as a prime source of American domestic and international power. Although the crisis of the century is passed, technology is essentially dehumanizing, Carleton says, and modern man's education must equip him to balance scientific with aesthetic and spiritual values.

Landmark Study of Historic Preservation.
Albert Rains, Laurance Henderson, *et al., With Heritage So Rich* (New York: Random House, 1966), 230 pp., illus., $10.00. This seminal report by a special national committee on historic preservation played a large role in the enactment of the National Historic Preservation Act of 1966, which defined the scope of national preservation policy and implemented several institutional arrangements designed to further historic preservation in America. "A nation can be a victim of amnesia. It can lose the memories of what it was, and thereby lose the sense of what it is or wants to be." (p. 1) This is the strongest rationale for expending public funds for the preservation and restoration of historic buildings and sites, argue the report's authors. The articles that make up the volume offer a portrait of the history and state of historic preservation in America, and make an emotional and powerful plea for the necessity of preserving man's visible tradition. The approach is somewhat absolutist in its values; little concern is shown for countering the equally insistent arguments of developers who speak in the language of efficiency and economics for the advisability of modernizing the facade of American cities.

Human Ecology and Building Design.
Robert Gutman, ed., *People and Buildings* (New York: Basic Books, 1972), 471 pp., $12.50. This collection of writings by biologists, anthropologists, sociologists, psychologists and architects analyzes the effects of working and living environments of human behavior. Collaboration between the disciplines of design and behavioral science has proven vital to the development of an architecture responsive to human needs. The editor has assembled material representing the convergence of social and design concerns in four basic areas: human behavioral constraints on building design, the impact of spatial organization on social interaction, the role of environmental influences on physical and mental health, and the significance of architecture as an expression of and reinforcer of social values.

Improving the Quality of Historic Preservation.
The Association for Preservation Technology, Box 2682, Ottawa 4, Ontario, Canada. The Association for Preservation Technology is a Canadian-American organization of professional preservationists, restoration architects, furnishings consultants, museum curators, architectural educators, archeologists, craftsmen, and other individuals involved in historic preservation. The Association serves as an information link among groups and individuals involved in preservation ac-

tivities and as a forum to promote the quality of professional practice in historic preservation. It publishes a quarterly *Bulletin* which deals with such issues as early roofing and masonry materials and practices, architectural histories of buildings and gardens, and historic landscape restoration. Further information may be obtained from Miss Meredith Sykes, Secretary-Treasurer.

Devising an Institutional Context for Environmental Protection.
Lynton K. Caldwell, *In Defense of Earth: International Protection of the Biosphere* (Indiana University Press, 1972), 295 pp., $8.50. Reviews the role of political organizations in safeguarding the biosphere, with particular reference to international cooperation.

Socio-Ethical Issues in Medicine.
Institute on Human Values in Medicine. Society for Health and Human Values, 825 Witherspoon Building, Philadelphia, Pennsylvania 19107. The Institute on Human Values in Medicine of the Society for Health and Human Values is designed to enhance teaching about human values in relation to the study and practice of medicine and the health sciences. To this end the Institute has developed a three-pronged program for cross-disciplinary studies in medicine and the humanities: consultative services offer aid to institutions interested in developing programs in the medical humanities, providing suggestions on program technique and content; Institute fellowships provide for individual cross-disciplinary study in residential settings; liaison and information services aid in the coordination of existing and planned programs in the medical humanities. The work of the Institute is supported by grants from the National Endowment for the Humanities. The parent organization, the Society for Health and Human Values, serves to sponsor, encourage, and coordinate the consideration of human values in health research and services.

Transportation Technology and Social Needs.
Richard M. Zettel, "The Art of Transportation Planning, Circa 1970," working paper for Workshop on Social Directions for Technology of the National Academy of Engineering, July 1970, 34 pp., from the Institute of Transportation and Traffic Engineering, University of California at Berkeley. The report discusses transportation planning in the San Francisco Bay Area, providing an informative overview of the network of variables — social, political, economic, and geographic — that must be taken into account in the process of urban transportation planning. It considers the work of the Bay Area Transportation Study Commission, whose approach to planning reflects the belief of engineering educators that modern technology necessitates a move away from a narrow discipline approach to problem-solving. The Commission staff includes engineers, planners, economists, sociologists, statisticians, and generalists, in an attempt to "start from the problems themselves and think towards the methodology of solution rather than from the disciplines thinking toward their application to the problem."

176

Social Scientists in the Private Sector.
Matthew Radom, *The Social Scientist in American Industry* (Rutgers University Press, 1970), 210 pp., $7.50. Why do social scientists accept employment in industry? In what terms, with what satisfactions or dissatisfactions do they view their roles and careers in the private sector of the economy? Designed in part to provide industrial management with insights that will help industry utilize its social science manpower most effectively, the study suggests that a major key to such utilization is the maintenance of an industrial climate of "autonomy and discovery" in which the social scientist is assured economic security while being encouraged to focus his creative energies on large-scale research and professional endeavors. Neither larger salaries nor opportunities for management training appear to be significant motivators for social scientists in industry; the major motivator is the opportunity to work on the solutions to important problems which call on the social scientist's special training.

Although Radom's findings indicate that a minimal level of ideological agreement with the position of the industry must be maintained by the social scientist if he is to remain in the industrial setting, Radom's study fails to suggest any ways in which the values of the social scientist-as-humanist interact with the values of the private sector. The study fails to indicate, in particular, how the institutions of technology are themselves influenced by the presence of social scientists in the ranks of industry.

"The Magazine of Alternative Science and Technology."
Undercurrents in Science and Technology, quarterly, from the UNDERCURRENTS partnership, 34 Cholmley Gardens, Aldred Road, London N.W. 6, England, annual subscription $2.50. Issue number 3 (Autumn/Winter 1972) includes an "Alternative Technology Guide."

Toward a Wider Understanding of Social Aspects of Science.
American Association for the Advancement of Science, *Science for Society; a Bibliography*, annotated entries on more than 3,000 publications chosen with an eye to use by teachers of natural science and social studies at the high school level. The fourth edition (1973) is available from the Association, 1515 Massachusetts Avenue, N.W., Washington, D.C. 20005. ($1.00) Current news items on both science education and social policy aspects of science are published in *Science Education News*, bimonthly available by request.

Reconciling Fact and Value.
W.T. Jones, *The Sciences and the Humanities; Conflict and Reconciliation* (University of California Press, 1965), 282 pp., $6.50. Attributes the difference in aims between science and humanism to the character of the languages they employ rather than to essential differences of facts from values. Different values are "realizable" in each realm and they are connected by a "linguistic continuum."

Evolution of Contemporary Values.
Jacob Bronowski, *Science and Human Values* (N.Y.: Harper & Row, 1965), 119 pp., "Perennial Library" paperback, $1.25. Because science derives laws and philosophical unities from facts and social values are empirically derived from "what works" in man's experience, Bronowski argues that the domains of science and human values intersect and therefore would not constitute distinct institutional systems.

Human Values and Technology: A Compendium.
Maxwell H. Goldberg, *Needles, Burrs, and Bibliographies; Study Resources: Technological Change, Human Values, and the Humanities* (1969), Center for Continuing Liberal Education, Pennsylvania State University, vi + 200 pp., mimeo. Includes a list of 299 recent publications on the character and status of the humanities by James H. Stone and a very extensive bibliography on technology and human values.

General Value Theory: An Overview.
Kurt Baier and Nicholas Rescher, ed., "Bibliography on the Theory of Value," *Values and the Future* (New York: The Free Press, 1969), pp. 489-512. The principal topics treated are methodological questions, major Anglo-American writings on general value theory, scientific approaches to value, and other bibliographies. A compilation of immense scope and learning.

Technology and a Crisis in Values.
Lewis Mumford, *Art and Technics* (Columbia University Press, 1952), 162 pp., $6.00. "Salvation lies, not in the pragmatic adaptation of the human personality to the machine, but in the readaptation of the machine, itself a product of life's needs for order and organization, to the human personality." (p. 14) A prophetic, eloquent portrayal of the need for a basic readjustment of technology and human values.

History Revitalizing Humanity.
Lewis Mumford, *Interpretation and Forecasts, 1922-1972: Studies in Literature, History, Biography, Technics, and Contemporary Society* (N.Y.: Harcourt Brace Jovanovich, 1973), 522 pp., $12.95. A wide-ranging presentation summarizing Mumford's work on many topics, this collection of essays and other writings affords a general review of Mumford's critique of technology. The misuse of technology is blamed on personal attitudes rather than its institutional involvements.

The Setting for Values.
W. T. Jones, "World Views: Their Nature and Their Function," *Current Anthropology*, Vol. 13, no. 1 (Feb. '72), pp. 79-109. The role of culture in affording shared frameworks for values is reviewed in a stimulating exchange between the author and commentators.

178

Limits to Scientific Validation.
Paul Roubiczek, *Ethical Values in the Age of Science* (Cambridge University Press, 1969), 318 pp., $10.00. A satisfactory justification for morality and ethics, the author argues, cannot be found in history, psychology or sociology, nor in the realm of objective or scientific knowledge. He speaks to the significance of philosophical truths reached by a Kierkegaardian type of "subjective knowledge" even in a scientific society, and posits values of goodness, truth, and beauty.

The Cultural Contexts of Technology.
Harry Woolf, ed., *Science as a Cultural Force* (The Johns Hopkins Press, 1964), 110 pp., $6.00. This short volume is a compilation of five lectures concerning science and culture in their broadest possible contexts. "Science and the society which it helps to shape, as it is influenced by it inturn, constitute a single culture in whose broad sweep the dialogue and the vocabulary may vary dramatically, but the grammar of learning remains constant." (p. 3) By dividing the discussion into the parts of science that intersect with government, technology, philosophy, and the humanities, the essays attempt to bridge the gap in vocabulary here noted, and reveal the unity of science and society.

Science as a Cultural Agency.
Paul A. Weiss, *Within the Gates of Science and Beyond; Science and Its Cultural Commitments* (New York: Hafner Publishing Company, 1971), 328 pp., $9.95. "Science is decidedly not a neatly bounded compartment that could be sequestered from the rest of civilized society as such, to be either enthroned as ruler or excised as tumor; on the contrary, it had better be described as a special attribute, a property, a peculiar way of looking at the world and of acting accordingly; in short, as a pervasive and indelible aspect of culture." (p. ii) The purpose of the present collection of essays is to elaborate on this answer to the question of the nature of science. Neurobiologist Weiss defines by example, by analogy, by discussion of current debates arising among scientists looking at their own discipline and so provides the layman with a "diversified and many-sided exposure" to science.

The Social Impact of Technology; An Anthology.
Melvin Kranzberg and William H. Davenport, eds., *Technology and Culture; an Anthology* (New York: Schocken Books, 1972), 364 pp., $10.00. An interdisciplinary approach to the understanding of culture as it has been shaped by technology is offered in this anthology of papers from the Society for the History of Technology. The articles reflect an opinion that man has the knowledge and understanding necessary to control the forces of change brought about by technology, to "rehumanize" it. To control technology, man must first understand it: "the student and the layman should develop a historical perspective, learn the impact of the machine on events, appreciate the meaning of technological revolutions, perceive the various roles of technology in society, and become aware of its interplay with such topics as art, values, and international relations." (p. 19)

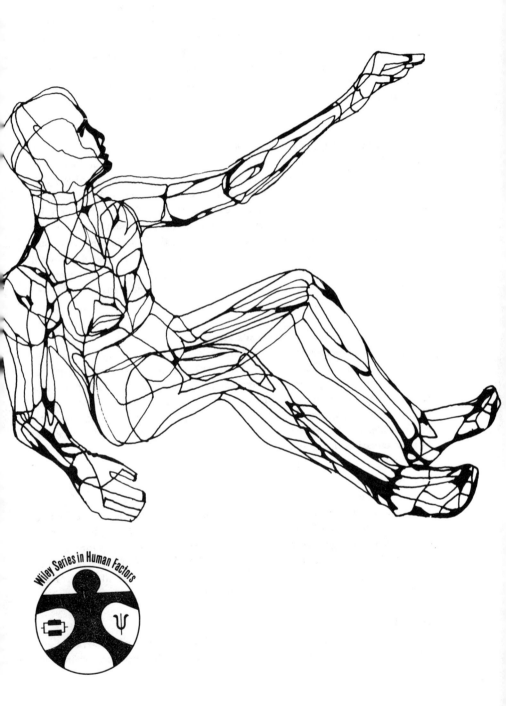

Equipment Design Constrained by Human Capabilities.
David Meister, "Wiley Series in Human Factors," *Human Factors; Theory and Practice* (New York: Wiley-Interscience, 1971), 412 pp., $17.25. The consideration of human factors in the future use of equipment has become widespread as a design practice. What is sought is maximum efficiency, calculated in terms of man-machine systems, in the way equipment will be operated. The role human factors specialists play in engineering organizations is described and their educational backgrounds, consisting primarily of psychology and engineering, are summarized.

Literature, Technology, and Applied Communication.
Edward M. Jennings, ed., *Science and Literature; New Lenses for Criticism* (Garden City, New York: Doubleday and Co., 1970), 262 pp., paper, $1.95. This collection of essays is intended to broaden the frame of reference of literary discussion; it is "interdisciplinary" in that the editor hopes to foster a sharing of interests and attitudes among the various disciplines of science and literature. The essays included span disciplinary concerns, including psychology, computer science, anthropology, and communications theory, and all are related to overall themes. Thus essays on information theory, probability, metaphor, and generative grammar are all concerned with man's capacity to create and transfer symbols — "the nature of verbal art."

Sources of Human Values: A Scientist's View.
Jean Rostand, *Humanly Possible; a Biologist's Notes on the Future of Mankind* (New York: Saturday Review Press, 1973), 182 pp., $6.95. Rostand's major interest is the uniqueness of man. He approaches a discussion of his subject through capsulization of his own thought and the thoughts of numerous other philosophers, poets, and scientists on the limits of the human experience, the origin of life, and the problems of biogenesis. The human qualities of science may be most accessible in the study of its history, he maintains.

The Debt of Science to Religion.
C.F. von Weizsacker, *The Relevance of Science; Creation and Cosmogony* (New York: Harper and Row, 1964), 192 pp., $5.00. To what extent is science the religion of the modern age? The author discusses his topic by tracing beliefs about creation and cosmogony from the myths of the ancient Middle East through current scientific interpretations of the origin and evolution of life. The book ends with a lengthy discussion of the increasing secularization of life and of the Christian roots of much in the scientific outlook.

The Cultural Role of Science.
Robert Bruce Lindsay, *The Role of Science in Civilization* (New York: Harper and Row, 1963), 318 pp., $6.50. This extensive inquiry into the role of science in civili-

zation begins with a definition of the fundamentals of scientific purpose and method and a partial refutation of the "alleged" dichotomy between the sciences and the humanities. The relationship of philosophy and history to science is briefly examined, with particular concern for the role of science in the development of human communication, information theory, and linguistic analysis. Only one chapter is devoted to a tracing of the relationship of technology to science, and a second to the links between technology and the State. In conclusion, the author considers the scientific study of human behavior and the problem of relating science to ethics.

Social Determinants of Scientific Values.
Roger G. Krohn, "Contributions to Sociology," Number 4, *The Social Shaping of Science; Institutions, Ideology, and Careers in Science* (Westport, Connecticut: Greenwood Publishing Corporation, 1971), 280 pp., $11.50. This behaviorist study of two groups of scientists in Minnesota attempts to determine recent changes in the "working conception" of science among scientists. Is scientific creativity being stifled, along with free research and basic research, due to the growth of organization and bureaucracy in science? How do attitudes toward science and their positions differ among scientists in universities, government, and industry? The findings indicate that scientists' own conceptions of science depend on the nature of their research support and the conditions of their research, which are in turn largely determined by government. Krohn concludes with a sociological scan of the self-attitudes of scientists and predicts the future composition of the scientific establishment.

Anthropology and the Roots of Social Values.
Alexander Alland, Jr., *The Human Imperative* (Columbia University Press, 1972), 185 pp., $2.95. This book is a defense of man against simplistic biological determinism. The author is a Darwinian and anthropologist, a believer in human variation, and does not share the pessimism of Morris, Lorenz, and Ardrey about the potential of man to salvage his problem-filled civilization. "Aggressive or passive behavior (and combinations of these) are both possibilities within the behavioral capacities of man . . . Human nature is largely open, and it is this very openness that gives the human species its great advantage." (pp. 23-24) The author addresses the same issues as the men he seeks to refute — the inheritance of aggression and territoriality — and his discussion assumes the dimensions of an analysis not only of biological man , but of his culture and recent history as well.

The Abdication of Value Inquiry by Social Science.
Ernest Becker, *The Lost Science of Man* (New York: George Braziller, 1971), 177 pp., $6.95. Becker treats the sociology of Albion Small and contemporary social anthropology in terms of their failure to provide a socially relevant science of man. Yet the development of a workable science of man, capable of providing an ideal model for society, is vital, he argues, to the future of the human race. "The best

we can hope for is to avoid the death and decay of mankind by using the feeble light of reason and the ideal of betterment." (p. xi) The search for a new "instrumental utopianism" must continue.

Cultural Evolution.
Hudson Hoagland, "Science and the New Humanism," *Science,* Vol. 143 (10 Jan 64), pp. 111-14. The characteristic that makes man unique among animals is his ability to direct and control his own evolution; science provides man with his most powerful tools for exercising that control. Hoagland speaks here of both biological and psychosocial or cultural evolution. Far from being a major obstacle to the evolution of new societal values, he argues, science is concerned with the very basis of change, the discovery of "truth," and Western man has long assumed that truth makes man free. In a scientifically-oriented society, excellence, independence, and originality are assets, and these factors, Hoagland feels, are necessary for the pursuit of truth. Furthermore, if accountability is a prerequisite for free action, science teaches man that he is free and responsible for his own actions. Thus science forms the basis, not for a loss of humane values, but for a new humanism.

Seeking to Define the Place of Values in the Scholarly Endeavor.
Max Black and June Goodfield, "Science, Technology, and the Humanities," *American Council of Learned Societies Newsletter,* Vol. 24, no. 2 (Spring 1973), pp. 7-29. Summaries of a discussion conference held in June of 1972. Are the humanities the moral and technology the instrumental side of the same scholarly ethos, as Carl Schorske maintained, or do the humanities have a domain uniquely their own? Leo Marx has proposed sustained colloquia of inquiry on such subjects as technology as a power tendency in society. June Goodfield pointed out that we have reached "an opportunity for revitalizing institutions which are in danger of having their patterns fossilized and their very lives frozen by tradition and by fear."

The Role of Art in Augmenting Humanism.
Edward Kamarck, ed., "The Humanist Alternative," *Arts in Society,* Vol. 10, no. 1 (Spring-Summer 1973), 151 pp., $2.50. This issue of *Arts in Society* is dedicated to a study of the relationship of art and humanism in modern society. The central essay in the collection takes the reader from the origins of modern art in the romantic period through a definition of humanism in modern art. Although we are encouraged to believe the opposite, the author argues, art is not "indifferent or independent of so-called non-visual values," and a humanism has long been reflected in an often-ignored "art that has expressed outrage at the castration of human life." (p. 11) Such humanist art is the antithesis of a second genre of modernist art which emulates science and technological invention and is viewed by most formalist art critics as the only valid modern art. Other essays further

define humanism and the relation of the artist to his audience and to his social vision.

Architecture and the Machine Aesthetic.
Reyner Banham, *Theory and Design in the First Machine Age* (New York: Praeger, 1960), 338 pp., $3.95. Banham presents a theory of architecture and sociology interpreting the overthrow of Ruskin's aestheticism and its replacement by a "machine age aesthetic" as a result of the deepening penetration of society by technologically advanced consumer goods. He considers how the symbolic program of modern architecture employs technological means to connote human values. As Mies van der Rohe said in 1928, "Both technology and industry face entirely new problems. It is very important for our culture and our society, as well as for technology and industry, to find good solutions." (p. 321)

Art, Technology, and Society.
Marcel Franciscono, *Walter Gropius and the Creation of the Bauhaus in Weimar; The Ideals and Artistic Theories of Its Founding Years* (The University of Illinois Press, 1971), 336 pp., $11.95. Franciscono's study concentrates on the early years of the Bauhaus and on the vision and goals of its founder Gropius in the first decades of the Twentieth Century. He views the Bauhaus as the "paradigm and culmination" of an entire era of German art, and thus is concerned with placing the Bauhaus within the context of German artistic development. Franciscono works with Gropius' unpublished papers and unstudied writings, and describes the often conflicting contributions of other artists such as Johannes Itten who were involved in the early Bauhaus.

Technological Values in Contemporary Art.
Jack Burnham, *Beyond Modern Sculpture; the Effects of Science and Technology on the Sculpture of this Century* (New York: George Braziller, 1968), 402 pp., illus., $15.00. In his scholarly overview of the influences responsible for modern sculpture, Burnham attempts to account for the shifting stylistic changes of the art form and identify both the underlying pattern and formal foundation of modern sculpture that might suggest future stylistic developments. Through an examination of Kinetic, Luminous, and Cyborg art, he traces the parallels between the development of modern sculpture and the intellectual and technical framework of modern science from their common roots in the philosophies of rationalism and materialism. The author notes, too, that a better understanding of modern materialism such as might be afforded by understanding its effects upon one art form, is vital to both artist and man if he is to control the directions in which materialism leads him.

Independence of Art, Science, and Technology.
John Adkins Richardson, *Modern Art and Scientific Thought* (University of Il-

184

linois Press, 1971), xix + 191 pp., illus., $10.00. Historical treatment of major art movements and attendant technological influences.

Art and Technology Merge in 1970 Pepsi Pavilion in Osaka.
Experiments in Art and Technology, *Pavilion,* (New York: E.P. Dutton and Co., Inc., 1972), 346 pp., illus., $6.95. *Pavilion* is the detailed account of the designing and construction of the Pepsi Cola Pavilion at the 1970 Expo in Osaka. The Pavilion was a work of art, an unprecedented collaboration involving over seventy-five engineers, artists, and industries in Japan and the United States, coordinated by a group of artists and engineers known as the Experiments in Art and Technology. The importance of the Pavilion as an example of close but open-ended cooperation among diverse group of designers, craftsmen, scientists, and businessmen was matched by its importance as a reflection of the change in artistic concern away from the art object and towards the individual's relationship to his environment. The Pavilion was meant to be experienced individually by each viewer; it was a "living, responsive environment", where fog was produced at will, moving floats responded to physical contact, and visual and auditory space were fluid and changing. The philosophy behind the Pavilion was expressed in the words of its coordinator: "The artist is a positive force in perceiving how technology can be translated to new environments to serve needs and provide variety and enrichment of life . . . . One of [the] objectives in relation to the Pavilion was to demonstrate physically the variety and multiplicity of experiences that the new technology can provide for the individual."

A New Design for the Humanities.
O.B. Hardison, Jr., *Toward Freedom & Dignity; the Humanities and the Idea of Humanity* (Johns Hopkins University Press, 1972), 163 pp., paper, $2.25. "Humanities departments are society's chief institutional means for providing students with a sense of human values and of the continuity of culture across the barriers of time and space. If their critics are right, they are doing the reverse." (p. 33) Hardison adduces the treatment of purpose in aesthetics as a model for humanistic discourse on values free of the coerciveness of ideology.

A Social History of Technology.
Herbert J. Muller. *The Children of Frankenstein: a Primer on Modern Technology and Human Values* (Indiana University Press, 1970), 431 pp., $10.00. This judicious review of the character and history of technology summarizes its impacts on twelve social realms such as war, business, the environment, and traditional culture. Defines technology broadly as "the elaborate development of standardized, efficient means to practical ends."

Civic Values and Social Traditions as Contexts for Technology.
Herbert J. Muller, *In Pursuit of Relevance* (Indiana University Press, 1971), 306

pp., $10.00. The author examines traditional study of the humanities and our traditional cultural heritage in general in an effort to distinguish what is relevant to life in today's society from what is not. He discusses ethics and freedom in the context of American politics and the need for a new moral code. He concludes with remarks about the significance of the counter culture and a statement of faith addressed to American youth.

Submission to Technique.
Jacques Ellul, *The Technological Society* (New York: Vintage Books, 1964), 449 pp., paper, $2.45. The neutrality of technologies in serving man is mythical, in Ellul's interpretation, because many technologies systematically converge on man resulting in "an operational totalitarianism." Eroticism and expressive music create only illusions of freedom, which can be attained in genuine form only by protecting the entire psyche from invasion.

The Human Side of Technology.
Eleanor Clark, *The Oysters of Locmariaquer* (New York: Vintage Books, 1966), 203 pp., paper, $1.65. Those who imagine a technology on a human scale, helping to offset rather than aggravate adverse environmental impact, will find in this book, an account of the artificial raising of oysters in Brittany, a description of a remarkable complex of techniques in their social and natural milieu.

Needed Reformulation of Cultural Symbolism.
Robert J. Lifton, "The Struggle for Cultural Rebirth," *Harper's Magazine,* Apr. 73, pp. 84-90. "Destruction and reconstruction — death and rebirth in the quest for immortalizing connectedness — is at the center of man's creation of culture. From this process alone can the urgently needed transformation of our own culture ensue." Thus Lifton summarizes his thesis that modern society's feelings of degeneration and loss are the harbinger of cultural rebirth. Western man has lost faith in the intricate web of images, institutions, and objects that comprise his culture, says Lifton; the technological age has produced massive uncertainties about man's present and future and has estranged man from his past. Man is therefore engaged, the article argues, in a search for "symbolic immortality" — i.e. "an expression of man's need for an inner sense of continuity with what has gone on before and what will go on after his own limited biological existence." Lifton believes that a significant aspect of this search is the current quest for "significant work experience" — a demand for harmony between man's immediate economic involvement and his social vision. New social arrangements are needed, he feels, to permit and encourage technology to become part of this cultural transformation. The author does not propose specific means for this transformation, but he does mention the "work commune" as a possible model, and he speaks of the need to allow human playfulness a place in the ideal work environment.

Recovering from Power Madness.
Robert Jay Lifton, *Home from the War; Vietnam Veterans: Neither Victims nor Executioners* (N.Y.: Simon and Schuster, 1973), 478 pp. $8.95. A cultural interpretation of the Indochina War as a theater for fraudulent, power-seeking institutions, portrayed in terms of the difficulty veterans of that conflict experience in finding genuine, socially redeeming involvements. The more insightful veterans readily recognized the limitations of conventional military technology in this setting. In the words of one veteran, "this giant technological element that we had was rendered impotent by a few little Vietnamese running around and throwing land mines here and there." (p. 171)

Institutionalizing Human Values in the Economic Order.
John D. Rockefeller 3rd, *The Second American Revolution; Some Personal Observations* (New York: Harper and Row, 1973), 189 pp., $6.50. Looking toward a person-centered rather than machine-centered society, the philanthropist-author asserts that the concerned and committed individual forms the backbone of a new American humanistic revolution. He argues that the challenges that various minority movements and scholar-critics have addressed to the capitalist system can be converted to important new forces of change if accommodated by "humanistic capitalism."

Institutionalizing Human Values.
Daniel Adelson, ed., "Community Psychology Series," Number 1, *Man as the Measure; the Crossroads* (New York: Behavioral Publications, 1972), 146 pp., paper, $3.95. Community psychology is a response to the need for a psychology in which "man is the measure"; social and institutional systems must be reexamined for their correspondance to man's needs in an explicitly value-laden context, and these systems must be changed if they need to be. "The shift is to a growth and development model, away from the traditional treatment model . . . from a concern essentially with mind-body problems . . . to a concern with a sound mind in a sound body in a sound community." (p. 15)

Man as Machine or Self.
J. Bronowski, *The Identity of Man* (Garden City, New York: The Natural History Press, 1965), 107 pp., $3.95. "The central theme of these essays is the crisis of confidence which springs from each man's wish to be a mind and a person, in the face of the nagging fear that he is a mechanism. The central question that I ask is: Can man be both a machine and a self?" (pp. 8-9) In answering the question that forms the theme of these essays, the scientist-poet-historian Bronowski defines the self and the machine in the light of recent discoveries in physics and biology, and in the light of literary and artistic "knowledge." Man, he finds, is "a machine by birth and a self by experience" and both parts of man — the mechanistic and the subjective — can be joined in a single realization of human dignity.

Technology and the Institution of Religion.
J. Edward Carothers, Margaret Mead, Daniel D. McCracken, and Roger L. Shinn, eds., *To Love or to Perish; the Technological Crisis and the Churches,* A Report of the U.S.A. Task Force on the Future of Mankind in a World of Science -Based Technology (New York: Friendship Press, 1972), 152 pp., paper, $1.95. The task force terms this collaborative essay a "manifesto for tomorrow," calling for an incorporation of Christian ethics into technological solutions to social problems. The writers call for a more active role for the churches in offering a "systematic and prophetic critique" of technological society and in shaping the images of a more perfect society. The essay honors the claim of religion to deal with the meaning of human life and define a role for humanism in improving the quality of life.

Displacement of Human Values by the Efficiency Principle.
Clarence J. Karier, *"Humanitas* and the Triumph of the Machine," *Journal of Aesthetic Education,* Vol. 3, no. 2 (April 69), pp. 11-28. Historical essay on the substitution of technological method for the transcendental values whose development had been the aim of liberal education in the Nineteenth Century.

Committed Science as a Source of Socially Destructive Tendencies.
Daniel Gasman, *The Scientific Origins of National Socialism; Social Darwinism in Ernst Haeckel and the German Monist League* (New York: American Elsevier, 1971), 208 pp., $13.50. Gasman's work is concerned with tracing certain key features of National Socialism back to nineteenth-century scientific positivism and materialism as revealed in the social Darwinism of the Monists and Ernst Haeckel, the German comparative anatomist. Monism's relationship to religion, political theories, and particularly theories of evolution, are explored at length.

Taking Society as the Context for Literary Studies.
Peter Brooks, "Romania and the Widening Gyre," *P.M.L.A. (Publications of the Modern Language Association of America),* Vol. 87 (1972), pp. 7-11. Argument for discarding tidy academic conventions restricting the experience of literature in order to treat it as a reflection of human society, technology, and values.

Participatory Design for Tomorrow's Cityscape.
OPTii (organization of planning teams international, inc.), 105 South Main Street, New Hope, Pennsylvania 18938, publishes *Noosphere,* 50c per issue. The organization is seeking to develop an "arcology" at a site still to be chosen.

Value Inquiry and Social Thought.
American Humanist Association, *The Humanist,* bimonthly, Paul Kurtz, Editor. Annual subscription $7.00, foreign, $1.00 additional. 923 Kensington Avenue, Buffalo, N.Y. 14215. *"The Humanist* attempts to serve as a bridge between

theoretical philosophical discussions and the practical applications of humanism to ethical and social problems . . . Ethical humanism sees man as a product of this world — of evolution and human history — and acknowledges no superhuman purposes." A very considerable publication featuring articles, editorials, book reviews, and coverage of the arts.

Values in Science and Religion: Bibliography.
"139 Significant Essays on Human Values in Relation to Science and Religion, 1966-1970," *Zygon: Journal of Religion and Science* (University of Chicago Press, 1970), 8 pp. A complete listing of authors and titles appearing in *Zygon* since its inception in 1966 through Volume 5 (1970). *Zygon* is a quarterly journal dedicated to the reunion of human values and science.

Behind the Facade of Objectivity.
Theodore Roszak, *Where the Wasteland Ends; Politics and Transcendence in Postindustrial Society* (Garden City, New York: Doubleday, 1972), 492 pp., $10.00. Technology is portrayed as culturally inadequate because the establishment of meaning turns upon symbols which find no counterpart in the pursuit of efficiency and objectivity. Roszak describes the art and thought of Blake, Wordsworth, and Goethe as manifestation of a more vital tradition which we risk psychic alienation and ecological disaster in disregarding.

Apocalyptic Humanism.
William Irwin Thompson, *At the Edge of History* (New York: Harper and Row, 1971), 180 pp., $6.95. "Getting Back to Things at M.I.T.," the third chapter of this book, stigmatizes the institution as the setting for Faustian bargains — students learn the techniques of power by suppressing consciousness. The principal aim of the book is to present an elaborate theory of the structure and evolution of institutions and a characterization of history as myth, in an attempt to clear a space for a drastically different re-imagining of the future.

Transcending Institutions.
William Irwin Thompson, "The Individual as Institution: The Example of Paolo Soleri," *Harper's Magazine*, Vol. 245, no. 1468 (Sept. '72), p. 48 ff. The author argues that private enterprise and private educational institutions have given way to a dominance of the public sector in modern America, coincident to a destructuring of society. Artistic capitalism in the hands of imaginative individuals has become a radical instrument for effecting cultural change. Soleri's attempts to redesign urban civilization utilizing arcology, a union of architecture and ecology, are viewed by the author as a consummation of the evolutionary process itself.

Conditioned Humans.
Rene Dubos, *A God Within* (New York: Charles Scribner's Sons, 1972), 325 pp., $8.95. Environmental dependence and capacities for communal life, having been impressed upon man through generations of primitive and pastoral existence, afford a basis for guiding the development of technology.

Toward a Humanistic Psychology.
Abraham H. Maslow, *The Farther Reaches of Human Nature* (New York: The Viking Press, 1971), 423 pp., $12.50. A collection of articles selected for publication by Maslow before his death in 1970. Among its major themes are the relationships of facts and values, an elaboration of concepts of personal growth, and the interplay of psychological and philosophical concerns.

Toward Adventures of Feeling.
George Steiner, *In Bluebeard's Castle; Some Notes Towards the Redefinition of Culture* (Yale University Press, 1971), 141 pp., $5.95. T. S. Eliot's *Notes Towards the Definition of Culture* omit scientific and technical developments. Steiner moves them to the center of the stage of consideration in a learned and graceful review. In Bluebeard's castle, the metaphor of the title, his last wife was driven to open each door until she confronted the ultimate and destructive truth. Is this the kind of commitment to knowledge that our culture is being asked to sustain?

Knowledge and Power.
Denis Donoghue, "God with Thunder," *Times Literary Supplement*, 3 Nov 72, pp. 1339-341. Compares the knowledge which led to man's expulsion from Eden with Prometheus's gift of practical knowledge. A powerful example of the treatment of value themes in literary criticism.

Value Change — A Humanistic Analysis.
Lionel Trilling, *Sincerity and Authenticity* (Harvard University Press, 1972), 188 pp., $7.95. In this historical interpretation the modern devotion to "authenticity" is shown to have emerged from the eighteenth-century literary value, "sincerity." A virtually unprecedented exegesis of the emergence and change of values as the primary historical process.

Institutional Criticism and Technology.
Arthur E. Morgan, *Dams and Other Disasters; a Century of the Army Corps of Engineers in Civil Works* (Boston, Mass.: Porter Argent, 1971), xxv + 422 pp., $7.50. A scathing criticism of the U.S. Corps of Engineers by the former Director of the Tennessee Valley Authority.

Technology in Its Institutional Matrix.
Harvey M. Sapolsky, *The Polaris System Development; Bureaucratic and Programmatic Success in Government* (Harvard University Press, 1972), xx + 262 pp., $9.95. A painstaking, detailed account of the circumstances governing the development and adoption of Polaris Missiles by the U.S. Navy.

Reinterpreting Economic Institutions
John Kenneth Galbraith, *The New Industrial State*, 2nd ed. (N.Y.: New American Library, 1972), Mentor paperback, xix + 404 pp., $1.95. Interprets technology as serving the cause of growth to which all large corporations are committed. Offers esthetic achievement as an alternative social goal, perhaps thus confounding the president of Union Oil quoted (on p. 331) as saying, "We should not fall prey to the beautification extremists who have no sense of economic reality."

Index

Abrams, M.H., 85.
Adelson, Daniel, 186.
Alchemy, teaching about, 11.
Alland, Alexander, 181.
American Society for Engineering Education, 152, 163.
Arendt, Hannah, 135, 137.
Art, science and engineering considered as, 19; response of scientists and engineers to, 28; science and technology as related to, 30-31; crafts and and technology, 109-126; ideology and culture, 127-141.
Association for Preservation Technology, 174.
Autobiography, scientific, in teaching, 13.
Averill, Lloyd J., 154.

Baier, Kurt, 177.
Banham, Reyner, 183.
Baram, Michael, 171, 172.
Barfield, Owen, 35.
Bartocha, Bodo, 161.
Bauhaus, 127-130.
Becker, Ernest, 181.
Beitz, Charles R., 100.
Bell, Daniel, 101.
Benn Anthony Wedgwood *et al.*, 172.
Berry, Wendell, 116.
Best, Fred, 169.
Billington, David, 155.
Black, Max, 83-93, 182.
Boorstin, Daniel, 169.
Bosserman, Phillip, 170.
Brademas, John, 155.
Braun, Werner von, 160.

Index

Brickman, William W., 170.
Bronowski, Jacob, 177, 186.
Brooks, Peter, 187.
Brown, Norman O., 120.
Brown, Sanford C., 170.
Bureaucracy, creation of its own truth by, 97.
Burke, John G., 159.
Burnham, Jack, 183.

Caldwell, Lynton K., 175.
Carleton, William G., 173.
Carnegie, Andrew, 115.
Carothers, J. Edward, *et al.*, 182.
Carr, E.H., 65-66.
Case Western Reserve University, 144.
Cetron, Marvin J., 161.
Citizen participation, essential to control social consequences of technology, 64; demands for, 76; as gimmickry, 81; minimized in military technology, 91-100; and the engaged intelligence, 133; mass movements, uncritical quality of, 137.
City, as formed by eight inventions, 106.
Clark, Eleanor, 185.
Coates, Vary T., 47-58, 161.
Columbia University, 148.
Commission on M.I.T. Education, 32; 39-46.
Communication with citizens on technical matters, sender-oriented reports as barriers to, 22; importance for technology assessment, 58; necessary to draw on socially dispersed insights, 62; superior to professional education in controlling technology,

71; black box theory, opinion input leads to policy output, 76.
Compton, Karl, 40.
Concourse, at M.I.T., 37, 150.
Congress, 51-52, 94.
Consumer goods and social sanctions of pleasure, 112.
Cornell University, 149.
Corporations, and technology assessment, 56-58; and military technology, 95-100; lack of self-correcting innovations in, 106-7; and crafts, 109-126.
Council for Science and Society, 151.
Crafts, 109-126.
Crane, R.S., 84.
Crowther, J.G., 172.
Curriculum, combining science, technology, and the humanities, 9-14; history of, 15; fragmentation in, 17; first two years, 17; pluralism in, 18; humanities requirements, 18; lack of time in, 18, 22; insufficiency of narrow excellence, 27; integration of science, technology, and the humanities in, 28-31; integration of knowledge in, 32; unfolding of consciousness in, 36; discouragement of humanities by, 39; engineering concepts, 81; needs for change in high schools, 107.

Dada, 127-28, 130.
Daddario, Emilio, 51.
Davenport, William H., 178.
Denny, Brewster C., 71-82.
Design, 134.
Dewey, John, 133.

Disaster inoculation, 74.
Disciplines, integration of, 29; innova
tions transcending, 107.
Discontinuity, age of, 102.
Discoveries, simultaneous, teaching of, 10.
Dominick, Peter, 168.
Donoghue, Denis, 189.
Douglas, Richard M., 15-19.
Draper, C.S., 163.
Drucker, Peter, 101.
Dubos, Rene, 32, 117-18, 189.

Eaton, Allen, 116.
Eberhard, John P., 47, 101-8.
Education, interaction of technological and humanistic aspects, 9-14; social aims and individual development, 15-19; technical competence too commanding a goal of, 40; of citizens, high priority f, 79; for significant innovation, 107.
Eliade, Mircea, 121-22.
Eliot, T.S., 109.
Ellul, Jacques, 37, 111, 185.
Engineering, education, 15-46; defined, 20; humanizing, 71.
Engineering Concepts Curriculum Project, 81.
Engineering Education, 152.
Engineers Joint Council, Technology Assessment Panel, 1144.
Environmental impact statements, 54.
Environmental quality task force, M.I.T., 46.
Erasmus, D., 85.

Erikson, Erik, 120.
Ernst, Morris L., 171.
Etzioni, Amitai, 160.
Ewald, William, 162.
Environmental study group, at M.I.T., 37.
Experiments in Art and Technology, Inc., 184.
Expressionism, 127-28, 134, and Nazism, 136.

Faith-Man-Nature Group, 148.
Federation of Americans Supporting Science and Technology, 167.
Feibleman, James K. 67-69.
Feldman, Stanley C., 67-69.
Firnberg, Hertha, 167.
Franciscono, Marcel, 183.
Frankel, Charles, 16, 157-59.
Freud, Sigmund, 128, 132.
Fromm, Erich, 171.
Full Employment Act of 1946, 74.
Fuller, Buckminster, 139.
Futures research, 59.
The Futurist, 163.

Gabor, Dennis, 163.
Galbraith, John Kenneth, 94-100, 190.
Gallie, W.B., 83.
Gasman, Daniel, 187.
Georgetown University, 145.
Gibbons, Michael, 59-66.
Giedion, Siegfried, 169.
Goheen, Robert F., 154.
Goldberg, Maxwell H., 156, 177.

Index

Goodfield, June, 182.
Green, Martin, 9-14, 146.
Gropius, Walter, 127.
Gutman, Robert, 174.

Habermas, Jurgen, 37.
Hamilton, Alexander, 115.
Hammond Report, 18.
Hardin, Tim, 109.
Hardison, O.B., Jr., 184.
Hartmann, Eduard von, 132.
Heisenberg, Werner, 29.
Henderson, Laurence, 174.
Henn, T.R., 156.
Henry, Jules, 114.
Herman, Theodore, 100.
Hill, Forest G., 166.
Hindle, Brooke, 24-28.
Hippocrates, 8.
Hirschfeld-Mack, Ludwig, 131.
History, planning of, 135.
Hoagland, Hudson, 182.
Hogg, Quintin, 160.
Holloman, J. Herbert, 168.
Horgan, J.D., 155.
Horowitz, Irving Louis, 165.
Human settlements, 101-8.
The Humanist, 187.
Humanitas, 84.
Humanities, integration with science and technology in instruction, 9-14; function at M.I.T., 15-46; service role, denial of, 15, 25, 32; intrinsic importance of, 16, compared to instrumental functions, 17; required courses in, 18; stolidity of, 23; im-portance of excellence in, 26; need to avoid vagueness, 29; isolation from science and technology, 32; inaudibility in technical institutions, 32; environmental depend-ence of, 39; marginality of at M.I.T., 39; institutional bias against at M.I.T., 40; inadequate impact on technology, 41; humanizing engineers, 71; analysis of, 83-83; as an "essentially contested" term, 83; need to create, 84; elements of traditional conception of, 84-85; and the Bauhaus program, 127.
Hume, Ivor, 115.
Husserl, Edmund, 33.
Huxley, Julian, 134.

Ideologies, defined, 69; problems of choice among, 69; involvement in technology, 72; and military requirements, 98; in art and technology, 127-141.
Impact of Science on Society, 161.
Institution on Man and Science, 149.
Institutions, inadequacies to problems, 22; false blame of, 24; M.I.T. as example of role in relating technology to human values, 15-46; environment in, 39; physical environment of, 40; favoring efficiency over reflectiveness, 41; need to transcend culture, 43; integrated with technology as-sessment, 60; and military technology, 94-100; efforts at self-preservation, 101; as needed for urban-oriented innovations, 106; challenge of crafts to, 109-126; frame-work for understanding, 127; and hope-

lessness, 139; multipurpose, to deal with values, 158-59.

Interagency Crafts Committee, 123, 126n.
International Society for Technology Assessment, 162.
Interpretation, difficulties of, 92.

Jennings, Edward M., 180.
Jones, Martin V., 162.
Jones, W.T., 176, 177.

Kamarck, Edward, 182.
Kandinsky, Wassily, 128-130, 137.
Kaplan, Max, 170.
Karier, Clarence J., 187.
Kasper, Raphael G., 162.
Kepes, Gyorgy, 131.
Klein, Richard E., 173.
Knowledge, narrowness of conventional concepts of, 33; crisis in conception of, 37; as based on imaginative participation in human experience, 92.
Kranzberg, Melvin, 178.
Krohn, Roger G., 181.
Kuhns, William, 171.

Laboratory directors, accomplishments, teaching of, 10.
Laird, Melvin, 95.
Lamm, Richard D., 173.
Language, 87-89.
Lehrer, Stanley, 170.
Lifton, Robert J., 185, 186.

Limits of the scientific disciplines, teaching of, 11.
Lindsay, Robert Bruce, 180.
Literature and science, teaching about, 12-14.
Lonergan, B., 62.
Loth, David, 171.
Lukacs, George, 136.

M.I.T., colloquium on the humanities in, 15-46; 71; 101; pattern set by research funding, 102; 150.
Making, sacral aspects of, 122.
Management techniques, threat to self-government, 75-76.
Manheim, Karl, 65, 132-33.
Marcuse, Herbert, 111, 120, 121.
Marland, William, 116.
Marx, Karl, 132.
Maslow, Abraham H., 189.
McCue, Gerald, 162.
McGowan, Alan, 153.
Meister, David, 180.
Metz, L. Daniel, 173.
Michaelis, Michael, 166.
Military Audit Commission, proposal for, 98.
Moholy-Nagy, Laszlo, 127-28, 133.
Montclair State College, 148.
Morgan, Arthur E., 189.
Muller, Herbert J., 184.
Mumford, Lewis, 177.

National Academy of Engineering, 50, 153.

196

Index

National Academy of Sciences, 50.
National Bureau of Standards, 146.
National Endowment for the Humanities, 146.
National Science Foundation, 146.
Nazis and ideological aspects of art and technology, 134-39.
Nelkin, Dorothy, 22.
Nolde, Emil, 136.
Novels, scientific, in teaching, 12-13.

Objectivity, n technology assessment, critique of, 60; in distinguishing science from the humanities, 85-93; and Karl Mannheim, 140n.
Occult tradition, teaching about, 11.
Office of Technology Assessment, 51-52, limits of, 65.
Operations research, compared to technology assessment, 59.
Order, good of, 61-62, 132.
Organization of Planning Teams International, Inc., 187.
Organization of science as a subject for teaching, 9-10.

Packard, David, 95.
Perez, Vincent, 7.
Perspectives, 86.
Phenomenology, 33.
Philosophy, foundational, 33.
Polanyi, Michael, 44.
Politics, necessarily involved with technology, 73.
Polytechnics, 151.

Post-industrial society, 102.
Princeton University, 147.
Pritchett, Henry, 40.
Prometheus, 159.
Psalm, 135th, 33.
Psychology and aesthetics, 131.
Psychology of science, teaching about, 11.
Public policy, four major needs of, 77.
Public-private distinction in the economy, implications for technology, 73-76; as applied to military suppliers, 99.

Radom, Matthew, 176.
Rains, Albert, 174.
Rescher, Nicholas, 177.
Richardson, John Adkins, 127-141, 183.
Rivers, Mendel, 97.
Rockefeller, John D., 3rd, 186.
Rogers, William B., 15, 40.
Rose, David J., 19-23.
Rostand, Jean, 180.
Roszak, Theodore, 188.
Roubiczek, Paul, 178.
Roush, J. Edward, 168.
Russell, Bertrand, 133.
Russell, Richard, 98.

Sapolsky, Harvey, 190.
Schaefer, Christopher, 32-38.
Schon, Don, 101.
Schultze, Charles, 98.
Schwartz, Eugene S., 170.
Science, and meaning of world, 35; compared to humanities, 83-93.

Science, popularization of, teaching about, 13.
Science fiction, teaching about, 12.
Science for Society: A Bibliography, 176.
Sherman Antitrust Act, 74.
Snow, C.P., 160.
Social problems, inadequacies of institutions, 22; narrowness of solutions a result of overspecialization, 23; interrelationships in curriculum, 36; technical problem-solving abilities, 40; byproducts of advances in science and technology, 43-44; role of technology in solving, 45-46; interest of business in solving, 56; to be overcome through social decision rather than by technology, 72; and pork barrel politics, 81; cities, 102; and technocracy, 132; similarities over time, 139.
Society for Health and Human Values, 175.
Sociology of Knowledge, 132-33, 137, 139.
Spark, 165.
Steiner, George, 189.
Steiner, Rudolf, 33.
Surrealism, 127, 130.
Systems analysis, 59.

Taylor, Frederick W., 111.
Technological forecasting, 59.
Technology, as art, 19; used by engineering in service to society, 20; social character of problems n, 20; need for communication with laymen, 22; history of, 24-28; and social change, 48-49, 57; distinguished from hardware, 49, 157; critique of treatment as independent variable, 63; as a subdivision of ethics, 67; ways of humanizing, 71; involvement with politics, 73; intervention in everyday life, 73; dangers to political system, 76-77; introduction into society, new institutions to regulate, 81; and military requirements, 94-100; cities and inventions characteristic of them, 105-8; and the craftsman, 109-126; and the Bauhaus, 127.
Technology assessment, teaching of, 22-23; institutional analysis of, 47-58; defined, 47; benefits from, 49-50; performers of, 53; case studies of, 53-54; and business, 56; and operations research, 59; need for integration with social decision-making, 60; need to consider socially dispersed insights and attitudes, 63; technology assessment system, 64; utopian qualities of, 66; international society for, 162.
Technology Review, 161.
Teller, Edward, 98.
Thompson, William Irwin, 188.
Thoreau, Henry David, 110, 111, 113.
Tools, 112-13.
Townsend, John, 110.
Tribe, Laurence H., 173.
Triggs, Oscar L., 115.
Trilling, Lionel, 189.

Undercurrents in Science and Technology, 176.
Unified Science Study Program, at M.I.T., 37.

Index

University of California, Berkeley, 142.
University of Colorado, 143.
University of Southern California, 142.
University of Wisconsin, Milwaukee, 144.

Values, humane, understanding of in higher education, 17; fostering consciousness of in relation to technology, 22; linkages to facts, education about, 36; and institutional environment, 39-40; deeper engagement with, in need of institutional encouragement, 41; rhythms of inner lives, 41; need for casualness in learning, 42; proposal for conference on knowledge and, 43; inadequacy of value-free methodologies in technology assessment, 60; and the "good of order," 61, 65; in the sociology of knowledge, 65, 158; bound up with aking and use of artifacts, 67; neutrality of the state regarding, 68; and ideology, 69; to be pursued through public processes rather than education of engineers, 71; and personal cognitive frameworks, 87; neutralization in science, 87; and language, 88; and the tasks of criticism, 88; dehumanizing effects of consumer economy, 110; and multipurpose institutions, 159.

Vanderbilt University, 145.
Voters, responsibility of, 79.

Warne, Aaron W. et al., 159.
Washington University, 143.
Watkins, C. Malcolm, 115.
Watson, James, 10.
Weber, Max, 37-38.
Wechsler, Judith, 38-31.
Weidenbaum, Murray, 96.
Weizsacker, C.F. von, 180.
Weisskopf, Victor, 44
Weizenbaum, Joseph, 32.
Whisnant, David E., 109-126.
Wirth, Louis, 65.
Woolf, Harry, 178.
Worcester Polytechnic Institute, 150.
Work, theories of, 111-14.
Work in America, 170.
Work-study, 36, 37, 123.

Yarmolinsky, Adam, 165.
Yeats, William Butler, 38.

Zettel, Richard M., 175.
Zygon, 188.

(No subject index entries for bibliography or institutional program summaries)

Recommended For Your Library

Recommended For Your Library